T0222114

Lecture Notes in Computer Science

Lecture Notes in Computer Science

Edited by G. Goos and J. Hartmanis

165

Thomas F. Coleman

Large Sparse Numerical Optimization

Springer-Verlag
Berlin Heidelberg New York Tokyo 1984

Author

Thomas F. Coleman
Department of Computer Science, Cornell University
Ithaca, NY 14853, USA

CR Subject Classifications (1982): G.1.6, G.1.3

ISBN 3-540-12914-6 Springer-Verlag Berlin Heidelberg New York Tokyo
ISBN 0-387-12914-6 Springer-Verlag New York Heidelberg Berlin Tokyo

Library of Congress Cataloging in Publication Data. Coleman, Thomas F. (Thomas Frederick),
1950- Large sparse numerical optimization. (Lecture notes in computer science; 165) 1. Mathematical optimization–Data processing. 2. Sparse matrices–Data processing. I. Title. II. Series.
QA402.5.C543 1984 519 84-5300
ISBN 0-387-12914-6 (U.S.)

Printing and binding: Beltz Offsetdruck, Hemsbach/Bergstr.
2145/3140-543210

PREFACE

In the spring of 1983 I taught a graduate course entitled 'Large Sparse Numerical Optimization'. Each week I prepared a set of lecture notes to reflect the material being covered. This manuscript is the result. The purpose of the course was to discuss recent developments in the area in order to prepare a student to pursue research. I hope this manuscript will prove useful to others with a similar purpose in mind.

My class was comprised of graduate students from applied mathematics, computer science, and operations research. All students had strong mathematical backgrounds and were familiar with computational methods in mathematics. In addition, most students had some familiarity with numerical optimization, optimization theory and graph-theoretic concepts (or at least I assumed they did). The students were required to do all exercises listed at the end of each chapter. In addition, each student either presented a paper in the area (-I supplied a number of possibilities) or worked on one of the research problems listed in the notes. It was a fun course !

The development of algorithms for large sparse numerical optimization is currently a very active area of research in numerical analysis. The adaptation of efficient methods to the large sparse setting is proving to be a difficult and challenging task. Apparently, it is often impossible to preserve sparsity and attain other *desirable properties* simultaneously: algorithms must achieve a delicate compromise. An exciting facet of problems in this area is that a full understanding of all algorithmic considerations requires ideas and concepts from many disciplines: eg. linear algebra, real analysis, data structures, graph theory, optimization theory, numerical methods, ... One of my goals in the design of this course was to emphasize this rich diversity of thought.

Though my personal research is mostly concerned with nonlinear optimization, I thought it best to start the course with a fairly detailed discussion of large sparse linear problems. The reasons are obvious: Firstly, they represent important optimization problems in their own right - there are still many open questions. Secondly, nonlinear problems are often solved via a sequence of linear approximations and therefore it is imperative that the complexity of the 'easy' subproblems be fully appreciated. Chapter 1 is devoted to large square systems of linear equations, Chapter 2 discusses overdetermined linear systems and Chapter 3 deals with large linear programs. The flavour of the first three chapters is somewhat different from the remainder of the notes: Their primary purpose to to provide some linear background to the more research oriented nonlinear chapters. Hence ideas are covered more quickly, almost in the style of a survey. (Nevertheless, I do point out research directions and interesting possibilities from time to time.)

The heart of the course and this manuscript is represented by Chapters 4 and 5. They are concerned with unconstrained nonlinear problems: equations, least squares and unconstrained optimization problems are all discussed with respect to algorithms, theoretical developments and current software. The material is presented unevenly with emphasis on topics and ideas of particular interest to me. I do not apologize for this: These notes are meant to reflect a personal and timely view of recent and current research in a lively field. They are not meant to be a tidy and complete presentation of a mature subject.

Finally, in Chapter 6, some of the issues concerning large quadratic programs are examined. This chapter is even less complete than the others - the intent is to give the reader a taste of the complexity and nature of large scale QP problems. I hope to teach a sequel course, in the near future, that deals with large QP's in greater depth.

I would like to thank Cornell University and the Department of Computer Science in particular for the opportunity to devote my time to such a course. Thanks also to Charlie Van Loan for encouraging me to prepare the lecture notes, to Mike Todd for enduring my lectures and keeping me on my toes, and to Jorge More' for reading a preliminary draft and suggesting improvements (some of which I implemented). I am grateful to the department's VAX 780, IMAGEN laser printer, and TROFF for helping me prepare these notes 'on the fly'. Finally, a heap of love and thanks goes to Marianne for supporting me through a most demanding term.

January 1984

Table of Contents

Chapter 1   Large Sparse Systems of Linear Equations

In this chapter we consider methods for solving $Ax = b$ where A is a large sparse matrix of order n. We will usually assume that we are able to store the nonzeroes of A and their locations in memory.

There are two basic approaches - direct and iterative. A direct method usually involves a matrix factorization such as

$$PAQ = LU$$

where P and Q are permutation matrices. The solution $A^{-1}b$ is then retrieved by solving

$$Ly = Pb, \ U\bar{x} = y, \ x \leftarrow Q\bar{x}.$$

Usually the game is to find P and Q so that L and U have a modest amount of fill. However, sometimes the objective is to find P and Q so that L and U are 'structured'. Of course the preservation of stability often places further restrictions on P and Q.

Iterative methods usually require less storage than direct methods. Typically the nonzeroes of A, their locations, and several working n-vectors need be stored in memory. Theoretically an infinite sequence of approximations $x^{(k)}$ is generated however in practice the iteration is stopped when a sufficiently accurate approximation is obtained.

The literature describing different approaches to these two problems is very extensive. In these notes we will cover a few fundamentals and briefly describe several techniques.

1.1 Direct Methods For Symmetric Positive Definite Linear Systems

This section is strongly based on the book of George and Liu[1981]. The reader is encouraged to consult this book for details concerning data structures, implementation of methods and the software package SPARSPAK.

If A is a symmetric positive definite matrix then the Cholesky factorization, $A = LL^T$, can be computed stably. There is no need to pivot or exhange rows or columns for stability reasons. The eigensystem of A is equivalent to the eigensystem of PAP^T, where P is a permutation matrix. Hence, for this large class of problems we can choose P (or, equivalently, symmetrically re-order the rows and columns of A) with exactly one concern in mind - the sparsity of $\tilde{L}$, where $PAP^T = \tilde{L}\tilde{L}^T$. In addition, P can be chosen *before any numerical computation begins*, provided it is known that A is symmetric and positive definite (SPD). This freedom is not available to us when A is a general indefinite matrix.

It is important to note that the Cholesky factorization is unique.

Theorem 1.1 If A is an n by n symmetric positive definite matrix, it has a unique triangular factorization LL^T, where L is a lower triangular matrix with positive *diagonal entries*.

Proof: See George and Liu [1981], for example.

Let A be a SPD dense matrix. There are three common ways to stably compute the Cholesky factor L (outer product, inner product and border schemes). Each method requires $\frac{n^3}{6} + O(n^2)$ operations: the methods differ in the manner in which data is accessed. We will not go into detail here except to briefly outline one of the algorithms: the outer product method. This algorithm often proves to be a useful way of viewing Cholesky decomposition.

The first step of the algorithm yields the first column of L, $l_{\bullet 1}$, and modifies the $(n-1)$ by $(n-1)$ lower-right submatrix of A. In particular, if we partition $A (= A_0)$ as

$$A_0 = \begin{bmatrix} a_{11} & \bar{a}_{\bullet 1}^T \\ \bar{a}_{\bullet 1} & \bar{A}_1 \end{bmatrix}$$

then the first step of the outer-product algorithm is

$$l_{\bullet 1} \leftarrow \begin{bmatrix} \sqrt{a_{11}} \\ \dfrac{\bar{a}_{\bullet 1}}{\sqrt{a_{11}}} \end{bmatrix}, \quad A_1 \leftarrow \bar{A}_1 - \frac{\bar{a}_{\bullet 1} \bar{a}_{\bullet 1}^T}{a_{11}} \tag{1.0.1}$$

To obtain the kth column of L, the above procedure is applied to A_{k-1}.

It is now easy to see that a judicious choice of P can lead to a drastic reduction in fill in the corresponding Cholesky factor. For example if A has the following structure,

$$A = \begin{pmatrix} \times & \times & \times & \times & \times \\ \times & \times & & & \\ \times & & \times & & \\ \times & & & \times & \\ \times & & & & \times \end{pmatrix}$$

then clearly L is completely dense (assuming that there is no cancellation). However, if P is chosen so that PAP^T has the form

$$PAP^T = \begin{pmatrix} \times & & & & \times \\ & \times & & & \times \\ & & \times & & \times \\ & & & \times & \times \\ \times & \times & \times & \times & \times \end{pmatrix}$$

then L suffers no fill. In particular, the structure of L is just the structure of the lower diagonal half of PAP^T. Hence we are led to the following problem: Given the sparsity structure of a SPD matrix A, find a 'good' permutation matrix P. It turns out that we can model Cholesky decomposition on graphs and the graph model is a useful way to view the problem.

1.1.1 The Graph Model

If A is a symmetric matrix define $G(A)$ to be the *adjacency* graph of A: $G(A)$ has n nodes $v_1,...,v_n$ and edges (v_i,v_j) if and only if $a_{ij} \neq 0, i \neq j$. Notice that $G(PAP^T)$ has the same structure as $G(A)$ however the labelling of the nodes may be different. *That is, choosing a permutation matrix P is equivalent to choosing an ordering, π, of the nodes of the adjacency graph.* Also, the number of nonzero off-diagonal elements in A equals the number of edges in $G(A)$.

Let A_0 denote the matrix A. Following the outer product scheme, it is clear that Cholesky decomposition can be modelled on graphs as follows. The graph $G(A_i)$ can be obtained from the graph $G(A_{i-1})$:

(1) Add edges so that the neighbours of node v_i (in $G(A_{i-1})$) are pairwise adjacent,

(2) Remove node v_i and its' incident edges.

Under the assumption that no cancellation occurs, this rule will correctly derive the adjacency graph of A_i, for $i=1,...,n-1$. Moreover, provided we ignore multiple edges, the structure of $L+L^T$ is given by $\overline{G}(\pi)$, where the edges of $\overline{G}(\pi)$ are the edges of $G(A_i)$, $i=0,...,n-1$.

The problem of finding a suitable permutation matrix P can now be rephrased: find an ordering π of the nodes so that $\overline{G}(\pi)$ is sparse, or perhaps 'structured.'

1.1.2 The Minimum Degree Ordering Algorithm

A reasonable objective seems to be the following.

Minimum Fill Problem: Find an ordering π of the nodes of G so that $\overline{G}(\pi)$ has the fewest edges over all possible orderings.

Unfortunately, this problem is known to be NP-complete. A proof is given by Yannakakis [1981]. Therefore we must restrict our attention to fast heuristic algorithms which give acceptable answers in practise. The most popular such ordering scheme is the minimum degree ordering algorithm due to Markowitz [1957] and Tinney [1969]. A description of this method follows. Let G_0 be the *unlabelled* adjacency graph of A.

Minimum Degree Ordering Algorithm

For $i=1$ to n,

    (1) In the graph G_{i-1}, choose a node of minimum degree: label this node z_i.

    (2) Construct graph G_i: Add edges so that the neighbours of z_i (in G_{i-1}) are pairwise adjacent; remove node z_i and its' incident edges.

At the completion of this algorithm we have determined an ordering that tends to produce little fill. In addition, we have determined the exact locations of the fill (symbolic factorization). Note that if this algorithm is implemented as stated then the maximum storage required is unpredictable (due to fill edges). George and Liu [1980] have implemented an efficient implicit scheme that does not explicitly remember the fill edges: storage requirements are the nonzeroes of A plus several n-vectors. The time complexity is proportional to the number of nonzeroes of L.

Theoretically, there is very little known about this algorithm in terms of the fill that it will produce. In practise its' performance is often quite acceptable for problems of moderately large size. However, for problems with known special structure other techniques are often preferred. Moreover, it is sometimes more efficient, in the end, to approach the problem with a different objective: find an ordering π to give $\overline{G}(\pi)$ special structure.

1.1.3 Band and Envelope Schemes

If A is a symmetric matrix then the *bandwidth* of A is defined by

$$\beta(A) = \max\{|i-j| \mid a_{ij} \neq 0\}$$

and the *band* of A is

$$band(A) = \{\{i,j\} \mid 0 < i-j \leq \beta(A)\}.$$

An important observation is that $band(A) = band(L+L^T)$. Therefore it is possible to implement Cholesky decomposition so that the zeroes outside the band are ignored: A *band method* ignores zeroes outside the band and treats all elements within the band as nonzeroes. It is easy to see that Cholesky decomposition requires $\dfrac{\beta^2 n}{2} + O(\beta^3 + \beta n)$ operations; the additional number of operations required to retrieve $A^{-1}b$, by forward and back substitution, is $2\beta n + O(\beta^2 + n)$.

A band approach is very attractive when β is small. The data structures needed are simple; overhead costs are low. There are several high-quality band-solvers currently available (eg. LINPACK). Sometimes a matrix comes naturally in banded form - a band-solver can then be used directly. Otherwise it will be necessary to symmetrically permute the rows and columns of A so that the bandwidth is minimized: again we are restricted to heuristics since Garey, Graham, Johnson, and Knuth [1978] have established that the corresponding decision problem is NP-complete.

Unfortunately, many sparse structures cannot be permuted into a band form with a small bandwidth. (Consider the arrowhead example). Therefore, for many problems the band approach would be very inefficient.

A slightly more sophisticated way of exploiting sparsity is an *envelope* or *profile* method. Let $f_i(A)$ be the column subscript of the first nonzero component in row i of A and define $\beta_i(A) = i - f_i(A)$. The *envelope* of A, denoted by $env(A)$ is defined by

$$env(A) = \{\{i,j\} | f_i(A) \leq j < i\}.$$

or equivalently

$$env(A) = \{\{i,j\} | 0 < i - j \leq \beta_i(A)\}.$$

The *profile* or *envelope size* is given by

$$|env(A)| = \sum_{i=1}^{n} \beta_i(A).$$

A critical observation is that $env(A) = env(L + L^T)$. Therefore it is possible to implement Cholesky decomposition so that the zeroes outside the envelope are ignored: an envelope scheme ignore zeroes outside the envelope and treats all elements within the envelope as nonzeroes. An envelope scheme is roughly as simple to implement as a band method; however, the profile of a matrix can sometimes be dramatically less than the band (consider the arrowhead example). A band method never 'operates' on fewer matrix elements since $env(A) \subset band(A)$. An envelope scheme allows for Cholesky decomposition in time $\frac{1}{2}\sum \beta_i^2(A) + O(\sum \beta_i)$. George and Liu discuss implementation and data structure issues in Chapter 4 of their book.

1.1.3.1 The Reverse-Cuthill-McKee Algorithm

The most popular profile reduction algorithm is the RCM algorithm. It is based on the following objective: if x and y are neighbours in the adjacency graph of A then their labels should be 'close'. To follow is a description of the algorithm when applied to a connected graph - for an unconnected graph this algorithm can be applied to each component separately. (The algorithm is a heuristic of course: I am quite sure that the profile decision problem is NP-complete however I am unaware of a proof.)

(1) Determine a starting node x_1.
(2) For i =1 to n: Find all the unnumbered neighbours of node x_i and number them in increasing order of degree.
(3) Reverse the ordering.

The 'reverse' was added to the Cuthill-McKee [1969] algorithm by Alan George - it has been proven that it is never inferior. The complexity of the algorithm is $O(m |E|)$ where E is the set of edges and m is the maximum degree of any node in the graph. We assume that a linear insertion sorting algorithm is used.

The effectiveness of the RCM algorithm is sensitive to the choice of starting node. A reasonable starting node is a *peripheral node*.

Let $G = (X,E)$ be a connected graph. The *distance* between any two nodes x and y is the length of the shortest path between x and y. The *eccentricity* of a node x is

$$e(x) = \max\{d(x,y)|y\epsilon X\}$$

The *diameter* of G is then

$$\delta(G) = \max\{e(x)|x\epsilon X\}$$

or equivalently

$$\delta(G) = \max\{d(x,y)|x,y\epsilon X\}.$$

A node is said to be *peripheral* if $e(x) = \delta(G)$.

It is generally deemed too expensive $(O(n^2))$ to find a true peripheral node as a starting node for the RCM algorithm: a *pseudo-peripheral* node is usually found instead by a fast heuristic algorithm. We will describe a popular procedure due to Gibbs, Poole and Stockmeyer [1976].

The algorithm depends on a *rooted level structure* concept. Let $G(X,E)$ be a connected graph; if $Y\subset X$ define $adj(Y)$ to be the nodes in G which are not in Y but are adjacent to at least one node in Y. Given a node $x\epsilon X$, the level structure rooted at x is the *partitioning* $L(x)$ satisfying

$$L(x) = \{L_0(x),...,L_{e(x)}(x)\},$$

where

$$L_0(x) = \{x\}, \quad L_1(x) = Adj(L_0(x)),$$

and for $i=2,...,e(x)$,

$$L_i(x) = Adj(L_{i-1}(x)) - L_{i-2}(x).$$

The eccentricity $e(x)$ is called the *length* of $L(x)$ and the *width* $w(x)$ of $L(x)$ is defined by

$$w(x) = \max\{|L_i(x)| \mid 0\leq i \leq e(x)\}.$$

Pseudo-Peripheral Node Finding Algorithm
(1) Choose an arbitrary node r in X.
(2) Construct the level structure rooted at r.
(3) Choose a node x in $L_{e(x)}(r)$ of minimum degree.

(4) Construct the level structure rooted at x; if $e(x) > e(r)$
   set $r \leftarrow x$ and go to (3)
(5) The node x is a pseudo-peripheral node.

Note that each pass of this algorithm takes time $O(m)$, where m is the number of edges. (I am unaware of a useful bound on the number of passes.) Envelope schemes are widely used and are often quite efficient for moderately large problems. Their implementation is fairly straightforward and their overhead costs are low. For larger problems, more sophisticated schemes may prove more efficient. These methods include one-way and nested dissection as well as quotient tree methods. See George and Liu [1981] for a discussion of such methods.

Problems with a particular structure are sometimes best solved with a specially tailored technique. We will describe such an example in the next section. (In Section 1.1.7 I suggest how this approach might be generalized.)

1.1.4 Bordered Structures

In many applications the matrix A has a natural bordered structure:

$$A = \begin{pmatrix} B & V \\ V^T & C \end{pmatrix}$$

where B is large and sparse (and possibly structured), V is sparse, and C is small and dense. For example, B could be a tridiagonal matrix of order 900 and C is of order 100. The Cholesky factor can be correspondingly partitioned:

$$L = \begin{pmatrix} L_B & 0 \\ W^T & L_{\bar{C}} \end{pmatrix}$$

It is possible to efficiently solve $Ax = b$ without storing W in memory. Hence the stored fill will be confined to L_B and $L_{\bar{C}}$. Notice that L_B and $L_{\bar{C}}$ are the Cholesky factors of B and $\bar{C}(= C - V^T B^{-1} V)$ respectively. Since $W = L_B^{-1} V$ we can write

$$V^T B^{-1} V = V^T (L_B^{-1}(L_B^{-1} V)).$$

This asymmetric partitioning suggest that W need not be stored: We can shunt the columns of V along, one at a time, and form the corresponding column of $\bar{C}$.

Let us assume that B is an order m band matrix with bandwidth β, and C is a dense matrix of order p. Here is an algorithm to construct L_B and $L_{\bar{C}}$ without storing W.

Factor Algorithm
(1) Factor $B = L_B L_B^T$,
(2) For $i=1$ to p: Solve $By = v^i$, $\bar{c}^i \leftarrow c^i - V^T y$,

(3) Factor $\overline{C} = L_{\overline{C}}L_{\overline{C}}^T$.

It is easy to verify that the number of operations required is approximately $\frac{m\beta^2}{2} + 2p\beta m + \frac{p^3}{6}$. If we partition $b = (b^1, b^2)$ and $x = (x^1, x^2)$ to correspond to the partitioning of A, then we can solve for x in the following way.

Solve Algorithm
(1) Solve $Bt = b^1$, $\overline{b}^2 \leftarrow b^2 - V^T t$,
(2) Solve $\overline{C}x^2 = \overline{b}^2$, $\overline{b}^1 \leftarrow b^1 - Vx^2$,
(3) Solve $Bx^1 = \overline{b}^1$.

It can easily be verified that the correct answer is obtained and the work involved is approximately $4m\beta + p^2$.

We have demonstrated that a band-bordered system can be efficiently solved provided a dense Cholesky decomposition routine and a band Cholesky decomposition routine are available (eg. LINPACK). A more general moral is this: sometimes an 'unusual' sparse structure can be exploited by partitioning the matrix and solving sub-systems involving 'common' structures. It may be possible to use commonly available software in an imaginative and efficient way.

1.1.5 Concluding Remarks Concerning SPD Systems

Modularity is possible for SPD systems because it is not necessary to pivot to ensure the stability of the decomposition algorithm. Hence we have the independence of the following four tasks:

1. Find a suitable ordering.
2. Set up the storage scheme.
3. Compute L.
4. Perform forward and backward substitution.

Notice that steps 1 and 2 involve the structure of A only; steps 3 and 4 proceed with the numerical computations and the storage structure remains static. If several problems with the same matrix A but different right-hand-sides b need to be solved then only step 4 need be repeated. If several problems involving different matrices A with the same structure need be solved, then only steps 3 and 4 need be repeated.

For most efficent methods the complexity of the above four steps is bounded by $O(\gamma^2 n)$, $O(\gamma n)$, $O(\gamma^2 n)$ and $O(\gamma n)$ respectively, where γ is the maximum number of nonzeroes in any row or column of L. (Of course constants are important and there is a wide variance amongst the various methods.) Therefore, for sparse problems these methods are considered *linear*. It is interesting to note the observed differences in time for the four steps. For example, George and Liu

[1981, p.283] present results for the RCM algorithm on a collection of test problems. The ratio of times for the four steps is approximately 5:1:70:10. Clearly the factor step (3.) dominates. We will see a different story in the general unsymmetric case.

1.1.6 Exercises

1. Verify that the outer product scheme performs as advertised: it computes the Cholesky factor of a symmetric matrix A if and only if A is positive definite.

2. Let $\eta(L_{*i})$ be the number of nonzeroes in column i of L, the Cholesky factor of A. Prove that the number of operations to compute L is approximately
$$\frac{1}{2}\sum_{i=1}^{n}\eta(L_{*i})^2.$$

3. Define the ith frontwidth of A to be

$$\omega_i(A) = |\{k \mid k>i \text{ and } a_{kl} \neq 0 \text{ for some } l \leq i \}|.$$

   Prove that in an envelope scheme the number of operations to compute the Cholesky factor is approximately $\frac{1}{2}\sum_{i=1}^{n}\omega_i(A)^2$.

4. Prove that $L+L^T$ has a *full envelope* if $f_i(A) < i$ for $2 \leq i \leq n$.

5. Prove that if linear insertion is used then the complexity of the RCM algorithm is bounded by $O(m|E|)$ where m is the maximum degree of any node in the adjacency graph and E is the set of edges.

6. Assuming no cancellation, show that $L_{ij} \neq 0$, $i>j$, if and only if there is a path from x_i to x_j in the adjacency graph of A, exclusively through some of the nodes x_k, where $k \leq j$.

7. Verify that the *Solve Algorithm* in Section 1.1.4 does indeed solve $Ax=b$. Verify the stated operation costs of the *Factor* and *Solve* algorithms.

1.1.7 Research Problems

1. Propose and investigate a general ordering algorithm which defines a partitioning

$$A = \begin{bmatrix} B & V \\ V^T & C \end{bmatrix}$$

   where B has a small bandwidth (or possibly a small profile) and C is relatively small. A possibility is to first apply a band minimizing algorithm (or possibly a profile minimizing algorithm) and then continue to throw nodes into C provided a net reduction in storage requirements (or possibly total cost) is achieved.

2. Investigate the applicability of recursive ideas to Section 1.1.4. For example, suppose that we initially find a maximal set of independent nodes in the adjacency graph and we identify the matrix B with this set. Hence B is diagonal. Clearly there is no fill in W and therefore we can expect $\overline{C}(=C-W^T W)$ to be sparse. The factorization of $\overline{C}$ would proceed in a similar (recursive) way: identify a maximal set of independent nodes in the adjacency graph of $\overline{C}$, ...

1.1.8 References

Cuthill, E., McKee, J., [1969]. *Reducing the bandwidth of sparse symmetric matrices*, in Proc. 24th Nat. Conf. Assoc. Comput. Mach., ACM Publ.

Garey M.R.,Graham R.L,Johnson D.S., and Knuth D.E. [1978].*Complexity results for bandwidth minimization*, SIAM Journal on Applied Mathematics, 34,477-495

George, A. and Liu, J., [1981]. *Computer Solution of Large Sparse Positive Definite Systems*, Prentice-Hall, Englewood-Cliffs, N.J.

George, A. and Lui, J. [1980], *A fast implementation of the minimum degree algorithm using quotient graphs*, ACM Transactions on Mathematical Software, 3, 337-358.

Gibbs, N.E., Poole, W.G., Stockmeyer, P.K.[1976]. *An algorithm for reducing the bandwidth and profile of a sparse matrix*, SIAM Journal on Numerical Analysis, 13, 236-250.

Markowitz, H.M.,[1957]. *The elimination form of the inverse and its application to linear programming*, Management Science 3, 255-269.

Tinney, W.F. [1969]. *Comments on using sparsity techniques for power system problems*, in Sparse Matrix Proceedings, IBM Research Rept. RAI 3-12-69.

Yannakakis, M. [1981], *Computing minimum fill-in is NP-complete*, SIAM Journal on Algebraic and Discrete Methods, 1,77-79.

1.2 Direct Methods For General Systems

In this section we consider direct methods for the solution of the large sparse system of linear equations $Ax = b$ where A is a general, possibly indefinite matrix. The most popular approach is to find permutation matrices P and Q so that $PAQ = LU$, where L is unit lower triangular and U is upper triangular. In addition, it is vital that L and U are sparse *and* the decomposition is stable. It is not possible to clearly separate structural and numerical questions (as we did in the SPD case) since pivoting for stability necessitates performing the numerical decomposition. However, there is one kind of structural analysis that can be performed before any numerical computations: determination of the block triangular structure of A.

1.2.1 Block Upper Triangular Form

Suppose that A has the following structure:

$$A = \begin{pmatrix} A_{11} & A_{12} & A_{13} \\ 0 & A_{22} & A_{23} \\ 0 & 0 & A_{33} \end{pmatrix}$$

Then $Ax = b$ can be solved in the following way:

(1) Solve $A_{33}x^{(3)} = b^{(3)}$,
(2) Solve $A_{22}x^{(2)} = b^{(2)} - A_{23}x^{(3)}$,
(3) Solve $A_{11}x^{(1)} = b^{(1)} - A_{12}x^{(2)} - A_{13}x^{(3)}$,

where $x = (x^{(1)}, x^{(2)}, x^{(3)})^T$, and $b = (b^{(1)}, b^{(2)}, b^{(3)})$. Hence in general, if p is the number of diagonal blocks, it is only necessary to decompose

$$P_i A_{ii} Q_i = L_i U_i, \quad i = 1, 2, ..., p.$$

It is clearly worthwhile to permute the matrix A to block upper triangular form, if possible.

The most common strategy has 2 phases. Firstly, permutation matrices $\bar{P}$ and $\bar{Q}$ are found so that the matrix $\bar{P}A\bar{Q}$ has a zero-free diagonal. (In fact it is only necessary to find *either* $\bar{P}$ or $\bar{Q}$: if this cannot be done then A is singular). Phase 2 involves applying symmetric permutations, $\tilde{P}$ so that $\tilde{P}\bar{P}A\bar{Q}\tilde{P}^T$ is block upper triangular. Duff[1976], George and Gustavson[1980], and Howell [1976] have established that the block upper triangular form is essentially unique, regardless of the choice for $\bar{P}$ and $\bar{Q}$. That is, the number of blocks and the particular rows and columns in each block is fixed: However, the ordering of the blocks and the ordering within blocks may vary.

1.2.1.1 Phase One: Bipartite Matching

Consider the *bipartite graph* induced by A, $G_B(A) = (R,C,E)$, where R is a set of n nodes $r_1, r_2, ..., r_n$, corresponding to rows and C is a set of n nodes $c_1, c_2, ..., c_n$ corresponding to columns. There is an edge (r_i, c_j) in the set of edges E if and only if $a_{ij} \neq 0$. Note that the structure of $G_B(PAQ)$ is identical to the structure of $G_B(A)$ for any permutation matrices P and Q: the *labelling* of the nodes may differ. A *matching* is an *independent* set of edges $E_M \subset E$: that is, the number of nodes touched by edges in E_M is *exactly* $2 \cdot |E_M|$. A *maximum matching* is a set E_M^* of maximum cardinality. If every node in G_B is touched by an edge in E_M^* then E_M^* is said to be a *perfect matching*. The importance of perfect matchings is demonstrated by the following result, which is easy to prove (constructively).

Theorem 1.2 Permutation matrices P and Q exist such that PAQ has a zero-free diagonal if and only if $G_B(A)$ has a perfect matching.

There exist good algorithms to find maximum matchings in bipartite graphs. Hopcroft and Karp[1973] have developed an algorithm which has a worst case complexity bound of $O(n^{.5}\tau)$ where τ is the number of nonzeroes in A. However, Duff[1981] prefers an algorithm with a complexity bound of $O(n\tau)$: Duff claims that *in practise* this algorithm exhibits a $O(n+\tau)$ behaviour and is almost always superior to the Hopcroft-Karp algorithm.

1.2.1.2 Phase Two: Strong Components

If A is a matrix with *zero-free diagonal* then a *directed graph* $G_D(A) = (X,E)$ can be defined as follows. Associate with each diagonal element a_{ii} a node x_i. Let there be a directed edge $(x_i \rightarrow x_j)$ if and only if $a_{ij} \neq 0$, $i \neq j$. The structure of the graph $G_D(PAP^T)$ is equivalent to the structure of the graph $G_D(A)$: the labelling (ordering) of the nodes may differ. A directed graph is *strongly connected* if there is a directed path from each vertex to each other vertex. A directed graph which is not strongly connected can be divided in a unique way into maximal strongly connected subgraphs which are called *strong components*.

It is not difficult to see that the block triangular partitioning of a matrix A is induced by the strong component partitioning of the graph $G_D(A)$. Tarjan[1972] has given a $O(n + \tau)$ algorithm to find the strong components of a graph. Duff and Ried[1976] have proposed an implementation of this algorithm tailored to this matrix application.

1.2.2 Threshold Pivoting

Let us assume that our system cannot be reduced to block triangular form. The usual way of blending stability and sparsity demands is known as *threshold pivoting*. That is, a pivot row is chosen on sparsity grounds unless a threshold stability condition is violated.

If we partition A as

$$A = \begin{bmatrix} a_{11} & \bar{a}_{1\bullet}^T \\ \bar{a}_{\bullet 1} & \bar{A}_1 \end{bmatrix}$$

then the first step of Gaussian elimination, with partial (row) pivoting can be viewed as

(1) Interchange rows so that $|a_{11}| = \max\{|a_{i1}|, i=1,...,n\}$,

(2) $l_{\bullet 1} \leftarrow \begin{pmatrix} 1 \\ \dfrac{\bar{a}_{\bullet 1}}{a_{11}} \end{pmatrix}, \quad u_{1\bullet} \leftarrow a_{1\bullet}^T$.

(3) $A_1 \leftarrow \bar{A}_1 - (\dfrac{\bar{a}_{\bullet 1}}{a_{11}})\bar{a}_{1\bullet}^T$.

To obtain the kth column of L and k^{th} row of U, the above procedure is applied to A_{k-1}. Note that the sparsity of column k of L equals the sparsity of column k of A_{k-1}; the sparsity of row k of U equals the sparsity of row k of A_{k-1}. Loosely speaking, the sparsity of $\bar{A}_k$ is modified by the product of the sparsities of the kth row and column of A_{k-1}.

If the pivot element is chosen as in step (1) then the algorithm is stable (in practise) since there is very limited growth in element size. However, the sparsity of A_k might be disastrously affected. If we want to minimize fill then the Markowitz[1957] condition is reasonable: In the first step we find the nonzero a_{jk} in $A_0 (=A)$ with the smallest product of other nonzeroes in its row with other nonzeroes in its column. This element is then pivoted to the (1,1) position. Step k is defined in a similar (recursive) way: Note that in step k we search matrix A_{k-1}. However, if we blindly follow this strategy, the stability of the decomposition might be disastrously affected. A reasonable compromise is to employ the Markowitz strategy subject to the pivot element being sufficiently large. For example, in the first step a_{11} should satisfy

$$|a_{11}| \geq \mu \cdot \max\{|a_{i1}|, 1 \leq i \leq n\}$$

where μ is in the interval (0,1]. In this way the growth of element size is controlled to some degree and sparsity demands are also considered.

Unfortunately this scheme requires that the sparsity pattern of the whole matrix be available at every step. Hence we need a data structure that can hold and update the matrix as the elimination proceeds. This is certainly possible using linked lists for both rows and columns however it may be inefficient. Reid[1976] suggests the following compromise.

Firstly, order the columns before any computation begins. For example, the columns may be ordered by increasing number of nonzeroes. Then process one column of A at a time: perform a *row interchange* to bring into the pivot position (k,k) the element in column k with the fewest nonzeroes in row k, *in the original matrix*, subject to the threshold stability condition mentioned previously.

To see that this is possible, suppose that we have computed L_k and U_k where L_k is n by k and lower triangular and U_k is k by k and upper triangular and

$$L_k U_k = A_{[k]},$$

where $A_{[k]} = (a_{*1}, \cdots, a_{*k})$. We obtain L_{k+1} and U_{k+1} after applying the previous transformations (stored in L_k) to column $k+1$ of A.

Arithmetically, this form of Gaussian elimination is equivalent to the previous version however this implementation allows for simpler data structures. Experiments by Reid[1976] suggest that there is some increase in the amount of fill compared to the Markowitz strategy.

1.2.3 Concluding Remarks Concerning Indefinite Systems

If a sequence of linear problems, $A^k x^k = b^k$, $k=1,...,$ are to be solved where A^k is 'close' to A^{k-1}, then the ordering and storage allocation determined for A^i might be useable for A^j, $j > i$. Reid[1976] suggests using the previous ordering and storage allocation while monitoring potential error growth. If the growth in element size is extreme, then a new ordering must be determined. Therefore in this context there are three separate tasks:

1) Analyze - The orderings, fill positions, and storage scheme are
     determined as well as the LU decomposition.
2) Factor - The LU decomposition is determined using previously
     determined orderings and storage allocations (potential
     error growth is monitored).
3) Operate - Given the LU factorization, the solution $A^{-1}b$ is found.

Reid[1976] gives numerical results comparing the three tasks - the time ratios are approximately 40:5:1. Notice that this contrasts sharply with the SPD case where the ordering/storage allocation step was much cheaper than the Factor step.

Except for block upper triangular form, we have not discussed structural approaches, for the indefinite case, in these notes. As we observed in the SPD case, it is often most efficient to exploit structure if possible. In the indefinite case allowance must be made for pivoting - this often results in ' blurring' the structure of A. For example, if A is banded with bandwidths β_L and β_U (lower and upper bandwidths respectively) then L and U will have bandwidths at most β_L and $\beta_U + \beta_L$, respectively.

Two of the available software packages for large sparse indefinite problems are the Yale Sparse Matrix Package (Eisenstat et al [1977]), and the Harwell routines (Duff [1977]), which are also part of the NAG library. The first package assumes that A has already been suitably permuted so that L and U can be stably computed and are sparse, where $A = LU$. The Harwell package includes routines for permutation to block upper triangular form and threshold pivoting.

These notes have just touched on some of the important issues concerning sparse indefinite systems. For more information, I recommend the excellent survey paper by Duff[1977].

1.2.4 Exercises

1. Prove that $rank(A) \leq |E_M^*(G_B(A))|$. Give an example where the above inequality is strict.

2. Prove Theorem 1.2.

3. State a theorem involving $rank(A)$ and

$$max\{\,|nonz(diag(PAQ))|\,|P,Q \ are \ permutation \ matrices\,\},$$

where $nonz(*)$ is the set of nonzero entries.

4. Verify that the 2 versions of Gaussian elimination described in Section 1.2.2 are equivalent in terms of arithmetic cost (disregarding pivoting strategies).

5. One approach to the indefinite problem is to replace it with an equivalent SPD problem. For example, we could replace $Ax = b$ with $A^T A = A^T b$, and then apply Cholesky decomposition along with the methods of Section 1.1. The advantages should be clear but there are three apparent problems:

i) There may be loss of information when forming $A^T A$.
ii) The accuracy of the computed solution to $(A^T A)x = A^T b$
depends on $\chi^2(A)$ instead of $\chi(A)$. Therefore,
there may be a serious loss of information.
iii) The matrix $A^T A$ might be considerably less sparse than A.

The first two problems can be overcome in the following way. Do not explicitly compute $A^T A$ but rather decompose $A = QR$ and form $Q^T b$, where Q is orthogonal and R is upper triangular, with positive diagonal elements. The solution can then be obtained in the standard way: solve $Rx = Q^T b$. It is not necessary to store Q at any time. That is, Q is a product of a sequence of

Givens transformations: Each member of the sequence if formed and temporarily stored (in about four words). The Givens transformation is then applied to the partially transformed matrix and right-hand-side and then discarded. George and Heath[1980] show that this can be done using the storage locations set aside for R. Notice that $R^T = L$ where $A^T A = LL^T$.

The third problem cannot be overcome so easily. It is easy to construct an example where A is sparse but $A^T A$ is dense. Nevertheless, a suitably sparse R might be obtained by forming the *structure* of $A^T A$ and then determining P such that the Cholesky factor of $P(A^T A)P^T$ would be sparse. This permutation matrix P would then be a suitable column permutation matrix for A: decompose $AP^T = QR$, (Q is not stored).

One possible disadvantage of this scheme is that if another system needed to be solved with the same matrix A but a different right-hand-side then it would be necessary to retrieve Q somehow. To overcome this problem, we could replace $Ax = b$ with $AA^T y = b$ and $x \leftarrow A^T y$. We can then obtain $A^{-1}b$ by decomposing $A = \hat{L}Q$ and then solving

$$\hat{L}w = b, \quad \hat{L}^T y = w, \quad x \leftarrow A^T y.$$

Paige[1973] has proven that the computed solution is dependent on $\chi(A)$ and not $\chi^2(A)$. Note that we can find a 'good' row ordering for A by examining the *structure* of AA^T.

a) Prove that the Cholesky factor of $A^T A$ is R^T, where $A = QR$, Q is orthogonal and R is upper triangular with positive diagonal entries.

b) Let P be a permutation matrix. Suppose that $A = QR$ and $PA = Q_1 R_1$, where Q, Q_1 are orthogonal matrices and R, R_1 are upper triangular with positive diagonal entries. Prove that $R = R_1$.

c) Give an example where A is much sparser than $A^T A$. Give an example where $A^T A$ is much sparser than A.

d) Suggest an algorithm to form the *structure* of $A^T A$ in *linear* time under the assumption that A has at most $\gamma \ll n$ nonzeroes in any row or column. It will be necessary to make reference to the particular data structure that you have chosen. How much storage space do you need ?

e) Suppose A has the following structure:

$$A = \begin{pmatrix} x & x & x & x \\ 0 & x & 0 & 0 \\ 0 & 0 & x & 0 \\ 0 & 0 & 0 & x \end{pmatrix}$$

What is the adjacency graph of $A^T A$? What is the adjacency graph of $L + L^T$ according to the graph model described in Section 1.1 ? What is the QR decompositon of A ? What is the apparent contradiction and how do you resolve it (algebraically)?

1.2.5 Research Problems

1. Investigate (experimentally) the approaches outlined in exercise 5 above. Compare (total) storage and execution times with a sparse LU solver (especially in the situation when many problems with the same structure must be solved). Suggest a way of handling a few dense rows (or columns).

1.2.6 References

Duff, I.S. [1981], *On algorithms for obtaining a maximum transversal*, ACM Transactions on Mathematical Software, 7, 315-330.

Duff, I.S. [1977], *MA28 -A set of FORTRAN subroutines for sparse unsymmetric matrices* , Report R8730,A.R.E., Harwell, England.

Duff, I.S. [1976], *On permutations to block triangular form*, Harwell report CSS 27, Harwell, England.

Duff, I.S. [1977]. *A survey of sparse matrix research*, Proceedings of the IEEE, 65, 500-535.

Duff, I.S. and Reid, J.K. [1976]. *An implementation of Tarjan's algorithm for the block triangularization of a matrix*, ACM Transaction on Mthematical Software 4, 137-147.

Eisenstat, S.C., Gursky, M.C., Schultz, M.H., Sherman, A.H., *Yale sparse matrix package, II. The nonsymmetric codes*, Computer Science RR 114, Yale University.

George, A., and Gustavson, F. [1980]. *A new proof on permuting to block triangular form*, IBM report RC8238, Yorktown Heights, N.Y.

George, A. and Heath, M., [1980] *Solution of sparse linear least squares problems using Givens rotations*, Linear Algebra and Its Applications, 34, 69-83.

Hopcroft, J.E. and Karp, R.M. [1973]. *An $n^{2.5}$ algorithm for maximum matching in bipartite graphs*, SIAM J. Computing 2, 225-231.

Howell, T.D. [1976]. *Partitioning using PAQ* , in Sparse Matrix Computations, edited by Bunch,J.R. and Rose, D.J., Academic Press, New York.

Markowitz, H.M. [1957]. *The elimination form of the inverse and its application to linear programming*, Management Science 3, 255-269.

Paige, C.C. [1973], *An error analysis of a method for solving matrix equations*, Mathematics of Computation, 27, 355-359.

Reid, J.R. [1976]. *Solution of linear systems of equations: direct methods (general)*, Lecture Notes in Mathematics 572, Cpoenhagen.

Tarjan, R [1972]. *Depth first search and linear graph algorithms*, SIAM J. Computing, 1, 146-160.

1.3 Iterative Methods For Linear Systems

Most of the work in this area has been done on structured linear systems that arise in a partial differential equations context. Therefore, it is not clear how relevant some of the techniques and conclusions are to the general large-scale optimization world. Hence we will not discuss some of the acceleration refinements that have been developed recently: some of them depend on structural properties which may not be present in optimization problems.

For more information on the method of conjugate gradients and splitting methods see Hestenes [1980], Young[1971], Hageman and Young[1981], and Varga[1962]. In addition, Golub and Van Loan [1983] discuss the interesting relationship between conjugate gradient methods and Lanczos procedures.

1.3.1 The Method of Conjugate Gradients

The method of conjugate gradients is applicable to symmetric (semi-) positive definite systems. An unsymmetric problem can be attacked by considering AA^T or A^TA. It is possible to implement conjugate gradients on such systems without actually forming the matrix products. The rate of convergence will depend on $\chi^2(A)$, however. ($\chi(A)$ denotes the condition number of A.) See Axelsson[1980] for more information.

1.3.1.1 Motivation

To follow is an argument to motivate the method of conjugate gradients. It is loosely based on a more detailed discussion given by Golub and Van Loan [1983].

Consider the following quadratic minimization problem:

$$minimize \ q(x) := -b^T x + \frac{1}{2} x^T A x \qquad (1.3.1)$$

where A is symmetric positive definite. Notice that the negative gradient of q(x) is the residual vector $r(x) := b - Ax$. Therefore x^* solves (1.3.1) if and only if $Ax^* = b$; an algorithm to minimize $q(x)$ is also an algorithm to solve $Ax = b$.

Let us develop an iterative procedure to minimize $q(x)$. We will generate a sequence of directions p_k and a sequence of approximate solutions: $x_{k+1} \leftarrow x_k + \alpha_k p_k$, $k=0,1,\dots$. The initial approximation, x_0, is arbitrary, however in this development we will assume that $x_0 = 0$.

A reasonable goal is this. Let us try and choose steplengths α_k and linearly independent directions p_k so that x_{k+1} *minimizes* $q(x)$ *in the linear subspace* $<p_0, \dots, p_k>$. This will ensure finite termination in at most n steps. Now suppose that x_k minimizes $q(x)$ in $<p_0,\dots,p_{k-1}>$. Clearly it follows that

$$r_k^T p_i = 0, \quad i = 0,...,k-1.$$

If $x_{k+1} = x_k + \alpha p_k$ then $r_{k+1} = r_k - \alpha A p_k$: Therefore, if $p_i^T A p_k = 0$ for $i = 0,...,k-1$ then

$$r_{k+1}^T p_i = 0 \quad i = 0,...,k-1.$$

Moreover, if we choose the steplength α optimally, then $r_{k+1}^T p_k = 0$.

Summarizing, if the search direction p_k satisfies

$$p_k^T A p_i = 0, \quad i = 0,...,k-1,$$

then $x_{k+1} (= x_k + \alpha_k p_k)$ will minimize $q(x)$ in the subspace $<p_0,...,p_k>$, provided α_k is chosen to minimize $q(x_k + \alpha p_k)$.

Since r_k is the direction of steepest descent at x_k, it is reasonable to choose p_k as the vector in the orthogonal complement of $<Ap_0,...,Ap_{k-1}>$ which is closest to r_k. That is, let us choose p_k to be the orthogonal projection of r_k onto the subspace orthogonal to $<Ap_0,...,Ap_{k-1}>$. If the sequence $x_1,...,x_k$ is generated in this fashion (ie. choosing p_k as defined above and using optimal steplengths) then the following properties are easy to prove for $j = 1,...,k$:

$$r_j^T p_i = 0, \quad i = 0,...,j-1 \tag{1.3.2}$$

$$r_j^T r_i = 0, \quad i = 0,...,j-1 \tag{1.3.3}$$

$$<p_0,...,p_{j-1}> = <r_0,...,r_{j-1}> = <b,Ab,...,A^{j-1}b>. \tag{1.3.4}$$

It turns out that the desired p_k, as defined above, can be efficiently computed as a linear combination of the residual vector and the previous direction:

$$p_k = r_k + \beta_k p_{k-1} \tag{1.3.5}$$

where $\beta_k := \dfrac{-p_{k-1}^T A r_k}{p_{k-1}^T A p_{k-1}} = \dfrac{||r_k||^2}{||r_{k-1}||^2}$. This leads us to the following finite algorithm.

Hestenes and Steifel[1952] Algorithm

$x_0 \leftarrow 0, \; r_0 \leftarrow b, \; k \leftarrow 0$

while $(k \le n-1)$ and $(r_k \ne 0)$ do

$\quad \beta_k \leftarrow \dfrac{||r_k||^2}{||r_{k-1}||^2} \quad (\beta_0 \leftarrow 0)$

$\quad p_k \leftarrow r_k + \beta_k p_{k-1} \quad (p_0 \leftarrow r_0)$

$\quad \alpha_k \leftarrow \dfrac{||r_k||^2}{p_k^T A p_k}$

$\quad x_{k+1} \leftarrow x_k + \alpha_k p_k$

$\quad r_{k+1} \leftarrow r_k - \alpha_k A p_k$

$\quad k \leftarrow k + 1$

od

Observe that we can think of this algorithm as a method to diagonalize the matrix A. That is, if $\hat{P} = (p_0,...,p_{n-1})$ then the matrix $\hat{P}^T A \hat{P}$ is diagonal.

On paper this algorithm looks very attractive. Storage requirements are minimal: we need enough space to store the matrix A, the vectors p_k, x_k, r_k and a work vector w_k to store $A p_k$. The work per iteration is roughly one matrix-vector product; at most n iteration are needed. Note also that if the matrix A is naturally available as a product of matrices, say $A = B \cdot C \cdot D$, then there is no need to form the product. Finally, if A represents the Hessian matrix of a non-linear function it may be possible to obtain the vector $A p_k$ without forming the matrix A at all (eg. finite differences), thus further reducing space requirements.

Unfortunately, things are not always as they seem.

1.3.1.2 Practical Conjugate Gradients

The behaviour of the conjugate gradient algorithm in practise can be far from ideal: rounding errors lead to a loss of orthogonality among the residuals and the algorithm may not converge in a finite number of iterations. Therefore it is better to view the practical algorithm as an infinite procedure which is terminated when the residual is deemed small enough. Hence the guard to the loop in the above algorithm should be replaced with a condition like

$$\|r_k\| \geq \epsilon \|b\|$$

where ϵ is a positive constant greater than or equal to machine precision.

Sometimes this modified algorithm works very well; however, for poorly conditioned problems the convergence can be devastatingly slow. *Note that even n iterations may be considered too many* : the resulting algorithm is $O(n\tau)$, where τ is the number of nonzeroes of A. Therefore, it may be advisable to accelerate convergence by *pre-conditioning*. That is, if C is a nonsingular symmetric matrix such that $\bar{A} := C^{-1}AC^{-1}$ has improved condition then it may be advantageous to apply the method of conjugate gradients to the system $\bar{A}\bar{x} = \bar{b}$, where $\bar{x} = Cx$, $\bar{b} = C^{-1}b$. If we define $M := C^2$, $\bar{p}_k := Cp_k$, $z_k := C^{-1}\bar{r}_k$, $\bar{r}_k = C^{-1}r_k = b - Ax_k$, then the pre-conditioned conjugate gradient algorithm can be expressed as follows.

Pre-conditioned Conjugate Gradient Algorithm

$x_0 \leftarrow 0,\ r_0 \leftarrow b,\ k \leftarrow 0$
while ($\|r_k\| \geq \epsilon\|b\|$) do
   solve $Mz_k = r_k$
$\beta_k \leftarrow \dfrac{z_k^T r_k}{z_{k-1}^T r_{k-1}} \quad (\beta_0 \leftarrow 0)$
$p_k \leftarrow z_k + \beta_k p_{k-1} \quad (p_0 \leftarrow z_0)$
$\alpha_k \leftarrow \dfrac{z_k^T r_k}{p_k^T A p_k}$
$x_{k+1} \leftarrow x_k + \alpha_k p_k$

$$r_{k+1} \leftarrow r_k - \alpha_k A p_k$$
od

The positive define matrix M is called the pre-conditioner. The big question is how to choose M. Clearly the first requirement is that systems of the form $Mz = r$ can be easily solved. The other objective is that the matrix $\bar{A}$ is better conditioned than A. Two common pre-conditioners are

$$M = diag(a_{11}, \ldots, a_{nn})$$

and

$$M = (D + \omega A_L)D^{-1}(D + \omega A_L^T)$$

where A_L is the strictly lower triangular part of A. These matrices are associated with the Jacobi and SSOR methods which we will discuss in the next chapter.

Another common pre-conditioner is given by an *incomplete Cholesky factorization* of A (Manteuffal[1979]). The basic idea is that it is assumed, a priori, that the Cholesky factor of A has a particular (and nice) structure. For example, it may be assumed that the factor is banded or that $a_{ij} = o \Rightarrow l_{ij} = 0$, for $i > j$. The 'Cholesky factorization' is then carried out to obtain the pre-conditioner except that fill outside the assumed structure is suppressed. A third possibility, which arises in an optimization context, is a limited memory quasi-Newton approximation - we will discuss this in more detail in a later section.

In many cases, with an appropriate choice of pre-conditioner, the *PCCG* method converges rapidly in $\ll n$ iterations (say $\sqrt{n}$). Nevertheless, it is difficult to recommend a *general* pre-conditioning scheme.
Usually, the pre-conditioner is tuned to the problem at hand; most results and experiments pertain to common linear systems which arise in a partial differential equations setting.

1.3.2 Splitting Methods

A splitting method based on the separation $A = B + C$ is defined by

$$Bx_{k+1} = -Cx_k + b. \tag{1.3.6}$$

Therefore it is necessary that systems of the form $By = z$ are easy to solve. In addition, convergence of such a method is achieved only if B is sufficiently 'close' to A. The following theorem formalizes this notion.

Theorem 1.3.2 A splitting method converges for all starting vectors x_0 if and only if $\rho(I-B^{-1}A) < 1$, where $\rho(M)$ is the spectral radius of the matrix M.

A proof is given by Stoer and Bulirsch[1981], for example. In general it is difficult to determine, a priori, if $\rho(I-B^{-1}A) < 1$. In some important cases convergence is known : If A is strictly diagonally dominant then the Jacobi and Gauss-Seidel methods are convergent. Indeed the Gauss-Seidel algorithm is convergent if

A is SPD.

1.3.2.1 Jacobi, Richardson and Gauss-Seidel Iterations

Each of the standard methods is defined by different choices for B and C. The Richardson (RF) method is defined by $B = I$ and $C = A - I$; the Jacobi iteration is determined by $B = D$ and $C = A_L + A_U$, where A_L (A_U) is the strictly lower (upper) triangular part of A, and D is the main diagonal of A. If $B = A_L + D$ and $C = A_U$ then the Gauss-Seidel iteration is defined.

The three methods can also be written as $z_{k+1} \leftarrow z_k + p_k$ where the methods are distinguished by the choice of correction vector p_k:

(Richardson) $p_k = b - Az_k = r_k$,

(Jacobi) $Dp_k = b - Az_k = r_k$,

(Gauss-Seidel) $(A_L + D)p_k = b - Az_k = r_k$.

Observe that if A is SPD then solving $Az = b$ is equivalent to solving the quadratic minimization problem (1.3.1). In this light each of the methods described above is an approximate Newton method with Hessian approximations I, D, and $A_L + D$ respectively. An obvious generalization of these methods is to modify the correction by a scalar: $z_{k+1} \leftarrow z_k + \alpha p_k$. In view of the optimization problem (1.3.1) a reasonable value for the scalar α is given by the optimal step: $\alpha = \dfrac{r_k^T p_k}{p_k^T A p_k}$. Typically these methods are not accelerated in this fashion but instead the following 'relaxed' approach is taken.

1.3.2.2 Relaxation

We will consider 'relaxing' the Gauss-Seidel step. The idea is to introduce a relaxation parameter ω so that $A_L + \omega D$ is the 'best' possible approximation to A. That is, let us determine the linear combination of A_L and D that best approximates A and let this approximation define the splitting in the Gauss-Seidel procedure. The splitting is defined by $A = B + C$ where

$$B = \omega^{-1}(D + \omega A_L), \quad C = \omega^{-1}[(1-\omega)D - \omega A_U].$$

The iteration matrix is therefore

$$B^{-1}C = (D + \omega A_L)^{-1}[(1-\omega)D - \omega A_U].$$

The objective is to choose ω so that $\rho(B^{-1}C(\omega))$ is minimized. Unfortunately, it is usually not possible to compute the optimal ω unless A has very special structure (eg. the adjacency graph of A is bipartite); it is sometimes possible to efficiently estimate the optimal ω - particularily if a congutate gradient acceleration technique is used (eg. Concus, Golub, and O'Leary[1976]).

The relaxed Gauss-Seidel procedure is called the SOR method. If $\omega > 1$ then we have an *over-relaxation* condition and if $\omega < 1$ then we have *under-relaxation*. Kahan [1958] established that for the Gauss-Seidel splitting and any matrix A,

$$\rho(B^{-1}C) \geq |\omega - 1|.$$

Therefore we are only interested in choosing $\omega \epsilon (0,2)$. If A is SPD then for all $\omega \epsilon (0,2)$ it can be proven that $\rho(B^{-1}C(\omega)) < 1$. Hence, if A is SPD then SOR is convergent for all $\omega \epsilon (0,2)$.

1.3.2.3 Symmetric SOR (SSOR)

It seems somewhat unnatural to approximate a symmetric positive definite matrix A with an unsymmetric matrix $B(\omega)$, as is done in the SOR method. The symmetric version of SOR (SSOR) is defined by

$$[\omega(2-\omega)]^{-1}(D + \omega A_L)D^{-1}(D + \omega A_L^T)p_k = r_k, \quad x_{k+1} \leftarrow x_k + p_k$$

Note that we can view the matrix

$$\bar{L} := (D + \omega A_L)(\omega[2-\omega]D)^{-\frac{1}{2}}$$

as an approximation to the Cholesky factor of A. (This leads us to consider that $\bar{L}\bar{L}^T$ might be a suitable pre-conditioner for the method of conjugate gradients and, on the other hand, any 'incomplete' Cholesky factorization might be considered a potential symmetric splitting matrix.)

1.3.2.4 Block Splitting Methods

The splitting methods discussed in the previous sections all have block counterparts. For example, suppose that A is partitioned in the following way:

$$A = \begin{pmatrix} A_{11} & A_{12} & A_{13} \\ A_{21} & A_{22} & A_{23} \\ A_{31} & A_{32} & A_{33} \end{pmatrix}$$

The block Gauss-Seidel iteration is defined, for $i = 1,2,3$, by

$$A_{ii} x_k^{(i)} = b^{(i)} - \sum_{j=1}^{i-1} A_{ij} x_k^{(j)} - \sum_{j=i+1}^{n} A_{ij} x_k^{(j)}$$

where vectors b and x are suitably blocked. A relaxation parameter can be introduced as in Section 1.3.2.2.

Of course it is necessary that the diagonal blocks be easily 'invertible'. Conflicting with this demand is the observation that it is clearly advantageous, from a rate of convergence point of view, to have large diagonal blocks (consider the extreme cases).

1.3.3 Exercises

1. Prove properties $(1.3.2) \rightarrow (1.3.4)$.

2. Give an example to show that the Hestenes-Steifel algorithm may converge in exactly 2 iterations with $n > 2$.

3. Show that the (implicit) pre-conditioned conjugate gradient algorithm given in Section 1.3.1.2 is mathematically equivalent to an explicit version. Why might the implicit scheme be preferable?

4. Suppose that A is partitioned

$$A = (A_1, A_2, ..., A_p)$$

Let x be blocked, $x = (x^{(1)}, \ldots, x^{(p)})$, to conform with the partition of A. Let $x_k^{(i)}$ be the least squares solution to

$$A_i x_k^{(i)} \approx b - \sum_{j=1}^{i-1} A_j x_k^{(j)} - \sum_{j=i+1}^{n} A_j x_{k-1}^{(j)}, \quad i = 1, ..., p.$$

Show that if A is invertible then the iteration is convergent: $x_k \rightarrow A^{-1}b$. Describe, in detail, an implementation of this algorithm with the additional assumption that columns within a group are stucturally independent. That is, if a_i and a_j belong to A_k then

$$a_{rj} \neq 0 \implies a_{ri} = 0$$

5. Consider a cyclic algorithm similar to that described in question 4 except that a column may belong to more than 1 group (each column must belong to at least one group). Prove or disprove: If A is invertible then the algorithm is convergent.

1.3.4 Research Problems

1. Investigate the procedure outlined in questions 4 and 5 above. Consider a partition $A = (A_1, \ldots, A_t)$, where $A_i^T A_i$ is banded with bandwidth m, for $i = 1, ..., p$.

2. Techniques which mix direct and iterative philosophies need investigation. For example, the bordered structure described in section 1.1.4 leads to 2 systems of equations - the second system involves the matrix $\overline{C} = C - W^T W$. This system could be solved using an iterative technique without actually forming the matrix. The first sytstem is efficently solved directly if B is of small bandwidth. This is just one example of a mixing strategy - there are a host of possibilities.

1.3.5 References

Axelsson, O [1980]. *Conjugate gradient type methods for unsymmetric and inconsistent systems of linear equations*, Lin. Algebra and Applic., 29, 1-66

Concus P., Golub G., O'Leary D. [1976]. *A generalized conjugate gradient method for the numerical solution of elliptic partial differential equations*, in Sparse Matrix Computations, J. Bunch and D. Rose (eds), Academic Press, New York.

Golub G.H. and Van Loan C. [1983]. *Advanced Matrix Computations*, The Johns Hopkins Press, Baltimore, Maryland.

Hageman, L. and Young, D. [1981]. *Applied Iterative Methods*, Academic Press, New York.

Hestenes, M. and Steifel, E, [1952].*Methods of conjugate gradients for solving linear systems*, J. Res. Nat. Bur. Stand., 49, 409-436.

Hestenes, M. [1980], *Conjugate Direction Methods in Optimization*, Springer-Verlag, Berlin

Kahan, W., [1958]. *Gauss-Seidel methods of solving large systems of linear equations*, Doctoral thesis, U. of Toronto, Toronto, Canada.

Manteuffel, T.A. [1979]. *Shifted incomplete factorization*, in Sparse Matrix Proceedings, I. Duff and G. Stewart (eds), SIAM, Philadelphia.

Stoer, J, and Bulirsch, R. [1980]. *Introduction to Numerical Analysis*, Springer-Verlag, New York.

Varga, R.S [1962], *Matrix Iterative Analysis*,Printice-Hall, Englewood Cliffs, N.J.

Young, D., [1971], *Iterative Solution of Large Linear Systems*, Academic Press, New York.

Chapter 2   Large Sparse Linear Least Squares

In this chapter we consider the overdetermined linear least squares problem:

$$minimize \ \|Ax - b\|_2$$

It will be assumed that A is of full column rank. We will emphasize direct methods for sparse problems. Iterative methods are also important - the reader is referred to the discussions and references given in Chapter 1. In addition, the method of Paige and Saunders[1978,1982] is an important contribution.

2.1 Dense Least Squares and The QR Decomposition

The book by Lawson and Hanson[1974] is an excellent source of information on least squares problems and their solution. Their emphasis is not on large sparse problems although there is some limited discussion; there is a wealth of useful information concerning rank deficient problems. In these notes we will be concerned only with the full column rank case - in Section 2.6 (linear constraints) this may refer to the 'projected' problem.

Let A be an n by t matrix of rank t. The least squares solution to the system $Ax \approx b$ is given by $x_{LS} = (A^T A)^{-1} A^T b$. A stable method to compute x_{LS} is to decompose $A = QR$, where R is t by t and upper triangular and Q is an n by t matrix with orthonormal columns. (Typically Q is the product of a sequence of Givens or Householder transformations and the product is not explicitly computed.) The least squares solution is then computed by solving $Rx_{LS} = Q^T b$. Note that it is not necessary to save Q - the successive orthogonal transformations can be applied and then discarded. (Discarding Q may not be wise if several problems, which differ only in their right-hand-sides, must be solved.)

It is common to think of the reduction of A to upper triangular form as a column oriented procedure. A row oriented variant is possible and might be preferable in some cases. For example, suppose that A is n by 5 with $n \geq 5$, and suppose that we have processed the first three rows of A to yield

$$(Q')^T A = \begin{pmatrix} x & x & x & x & x \\ 0 & x & x & x & x \\ 0 & 0 & x & x & x \end{pmatrix}$$

Let us consider processing the fourth row. We can apply a sequence of Givens transformations involving rows (1,4), (2,4) and (3,4), in turn, so that

$$(Q'')^T A = \begin{pmatrix} x & x & x & x & x \\ 0 & x & x & x & x \\ 0 & 0 & x & x & x \\ 0 & 0 & 0 & x & x \end{pmatrix}$$

Clearly this idea can be repeatedly applied until all rows are processed.

2.2 The Method of George and Heath [1980]

We will be very brief here - some of the underlying ideas are discussed in Section 1.2.4 (question 5). Here is the algorithm.

George-Heath Algorithm
1. Determine the *structure* of $B = A^T A$.
2. Find a permutation matrix P so that $\bar{B} := PBP^T$ has a sparse Cholesky factor $\bar{R}^T$.
3. Apply a symbolic factorization algorithm to $\bar{B}$, generating a row oriented data structure for $\bar{R}$.
4. Compute $\bar{R}$ by processing the rows of AP^T, ($Q^T(AP^T) = \bar{R}$,) using Givens rotations - apply the same rotations to b to get $\bar{b}$.
5. Solve $\bar{R}\bar{x} = \bar{b}$, $x \leftarrow P^T \bar{x}$.

Clearly the method has many strong features. Firstly, the method is stable and the accuracy of the computed solution is dependent on $\chi(A)$, if the residual is sufficiently small. (Note: For large residual problems, the computed solution is dependent on $\chi^2(A)$, regardless of the method.) Secondly, the familiar symmetric re-ordering routines can be used in step 2 ; there is also the very attractive separation of symbolic and numeric tasks. It is not necessary to store A in memory at any time - A can be examined row by row. It is not necessary to store Q at any time - the successive Givens rotations can be discarded. Memory must be allocated only for the (hopefully) sparse matrix $\bar{R}$ and the adjacency graph of $A^T A$.

On the negative side, a few dense rows in A will completely destroy this scheme since $A^T A$ will be dense. (Heath[1982] has suggested some special techniques for handling a few dense rows.) Another potential drawback is that if another least squares problem needs to be solved - identical except for the right-hand-side - then Q must be retrieved in some way. A possibility is to multiply the new b with A^T - there is some potential loss of accuracy, however. Finally, the amount of work to process a row is dependent on the row ordering - care must be taken to ensure that the work involved does not become excessive. George and Ng[1983] have suggested a row ordering strategy using a nested dissection approach.

2.3 Block Angular Structure

Suppose that A can be partitioned $A = (B,C)$ where B is block diagonal with rectangular blocks $B_1,...,B_p$ (each block is of full column rank). Partition the column-block C to conform: $C = (C_1, \ldots, C_p)^T$. This form is of interest because it arises naturally in some applications and more generally, there are ordering algorithms designed to yield such a form. Weil and Kettler [1971] suggested a heuristic algorithm to produce this form. The method of nested dissection will also do the job (George, Golub, Heath and Plemmons[1981]). Once in this form, the least squares problem can be stably and efficiently solved.

Algorithm For Block Angular Least Squares Structure

1. Compute $Q_i^T[B_i,C_i,b_i] = \begin{bmatrix} R_i \ S_i \ c_i \\ 0 \ T_i \ d_i \end{bmatrix}$ $i=1,...,p$

   where R_i is upper triangular.
2. Solve *minimize* $\|Tx_{p+1} - d\|_2$, where $T := (T_1, \ldots, T_p)^T$,

   and $d := (d_1,...,d_p)^T$
3. $i \leftarrow 1$

   while $(i \leq p)$ do

      solve $R_i z_i = c_i - S_i x_{p+1}$

   od

This approach and the method of George and Heath are appealing because they are stable and there is the convenient separation of numeric and symbolic tasks. The following idea, due to Peters and Wilkinson[1970] and extended by Bjorck and Duff[1980], does not enjoy such a convenient separation of tasks: numerical pivoting is required for stability. However, there *may* be some sparsity advantage since the method is based on the LU factorization of A instead of an orthogonal factorization.

2.4 The Method of Peters and Wilkinson[1970]

Suppose that P is a permutation matrix, chosen on stability and sparsity grounds, so that

$$PA = LU$$

where L is unit lower trapezoidal and U is upper triangular. The least squares problem , *minimize* $\|Ax - b\|_2$, is equivalent to

$$\text{minimize} \|Ly - Pb\|_2, \quad Ux = y \tag{2.4.1}$$

It can be argued that L will typically be well-conditioned - any ill-conditioning present in A is reflected in U. Therefore, the linear least squares problem (2.4.1) can be reliably solved by the method of normal equations: we can expect a reasonably accurate solution since $\chi^2(L)$ will usually be 'small'.

2.5 Indirect Methods

The methods of section 1.3 are all applicable here - they will involve the matrix A^TA (implicitly) and the rate of convergence will depend on $\chi^2(A)$. The interested reader should also study the method of Paige and Saunders [1978,1982].

2.6 Linear Equality Constraints

We consider the problem

$$minimize \ \|Ax - b\|_2, \quad Cx = d \tag{2.6.1}$$

where A is m by n and C is t by n with $t \leq n \leq m$. We will assume that rank $(C) = t$.

2.6.1 The Projection Method

The basic philosophy behind this type of method is this. We first compute a point feasible to the constraints and then we restrict our computations to this linear manifold.

Let y_1 be a particular solution to the system $Cx = d$. For example, y_1 could be the least squares solution: $y_1 = C^T(CC^T)^{-1}d$. Every point feasible to the equations $Cx = d$ can be expressed as $y_1 + Z\bar{y}_2$ where Z is *any basis* for the null space of C. If we substitute this expression for x then (2.6.1) can be written as

$$minimize \ \| (AZ)\bar{y}_2 - (b - Ay_1)\|_2 \tag{2.6.2}$$

Notice that this problem has a unique solution if and only if AZ has rank $n-t$. The question we are faced with is how to choose Z. An orthonormal basis is preferable from a stability point of view since then the conditioning of AZ will not be worse than the conditioning of A. However, for large problems it is important that we can represent Z in a 'small' space. In addition we may want to explicitly form AZ - again for large problems it is necessary that AZ be small or sparse. Therefore it may be necessary to choose a basis that is not orthonormal but has better sparsity properties. To follow is a statement of a projection algorithm which employs sparse orthogonal principles to some degree, although we do allow for freedom in the selection of the basis Z.

The Algorithm
1. Decompose $C^T = Q_1 R$ (each Givens rotation can be discarded after it is applied)
2. Construct Z : This may be combined with step 1 if we identify Z with Q_2 where $Q = (Q_1, Q_2)$.
3. Determine y_1: Solve $R\tilde{y} = d$, $R^T\hat{y} = \tilde{y}$, $y_1 \leftarrow C^T\hat{y}$.
   (Note: If Q_1 has not been discarded then: $R^T\tilde{y} = d$, $y_1 \leftarrow Q_1\tilde{y}$.)
4. $\bar{b} \leftarrow b - Ay_1$.
5. Form AZ.
6. Solve $minimize \|(AZ)\bar{y}_2 - \bar{b}\|_2$, $y_2 \leftarrow Z\bar{y}_2$.
7. $x \leftarrow y_1 + y_2$.

Note that we may not want to form AZ explicitly. In this case we could perform step 6 using an iterative method.

Finally, an important observation is that if A represents a Jacobian matrix then a finite difference approximation strategy may be quite efficient since we need only estimate AZ and Ay_1. This could result in considerable savings if $n-t$ is small.

2.6.2 The Elimination Method

Partition the matrix $\begin{bmatrix} C \\ A \end{bmatrix}$ as follows:

$$\begin{bmatrix} C \\ A \end{bmatrix} = \begin{bmatrix} C_1 & C_2 \\ A_1 & A_2 \end{bmatrix}$$

where C_1 is t by t and nonsingular. (It may be necessary to permute columns to ensure this.) If we partition z to conform with the partition of C, then the constraint equation $Cz = d$ can be used to solve for z_1:

$$z_1 = C_1^{-1}(d - C_2 z_2).$$

Substituting this expression for z_1 in $\|Az - b\|$ yields a new unconstrained least squares problem:

$$minimize \ \|\overline{A}_2 z_2 - \overline{b}\|$$

where $\overline{A}_2 = A_2 - A_1 C_1^{-1} C_2$, and $\overline{b} = b - A_1 C_1^{-1} d$. *It is important to realize that the choice of C_1 will affect the conditioning of the problem.*

Indeed this method is a *projection method* with $Z = \begin{bmatrix} -C_1^{-1} C_2 \\ I \end{bmatrix}$ and $y_1 = \begin{bmatrix} C_1^{-1} d \\ 0 \end{bmatrix}$. If t is small then Z is sparse however it is 'unlikely' that AZ $(\overline{A}_2)$ will also be sparse (see exercises 6 & 7 , however). If t is large then the number of columns in Z will be small (though probably dense). The basis Z can be computed if C_1 has a sparse factorization. For example, column i of Z is $\begin{bmatrix} w_i \\ e_i \end{bmatrix}$ where e_i is a unit vector with a -1 in component i and

$$C_1 w_i = (C_2)_i.$$

2.6.3 The Method of Lagrange Multipliers

The efficiency of this method is dependent on the number of constraints being relatively small. In addition, *it is necessary that the matrix A be of full column rank*: therefore, this method is not as general as the previous two procedures. We follow Heath[1982] in our description of the implementation.

Consider the quadratic programming problem:

$$minimize \; \text{½} \; x^T(A^TA)x - (A^Tb)^Tx, \quad Cx = d \qquad (2.6.3)$$

The vector x is a solution to this constrained quadratic programming problem (and hence the constrained least squares problem) if and only if there exist a vector λ such that

$$\begin{bmatrix} A^TA & C^T \\ C & 0 \end{bmatrix} \begin{bmatrix} x \\ \lambda \end{bmatrix} = \begin{bmatrix} A^Tb \\ d \end{bmatrix}$$

This system can be solved in a block fashion:

$$(A^TA)y = A^Tb, \quad [C(A^TA)^{-1}C^T]\lambda = Cy - d$$

followed by

$$x \leftarrow y - (A^TA)^{-1}C^T\lambda$$

Heath pointed out that it is unnecessary to solve for λ (completely) and proposes the following implementation.

Algorithm for Constrained Least Squares
1. Decompose $A = QR$
2. Solve $Ry = c$
3. Solve $R^TK = C^T$,
4. Decompose $K = \tilde{Q}L^T$, $\tilde{Q}$ is n by t with orthogonal columns
5. $r \leftarrow d - Cy$
6. Solve $Ls = r$
7. $v \leftarrow \tilde{Q}s$
8. Solve $Rz = v$
9. $x \leftarrow y + z$

If A is of full rank and C is small then this approach is quite attractive for large problems. Step 1 must involve column ordering, of course, to limit fill in R.

Another way to interpret the Lagrange multiplier approach is this: The unconstrained solution $((A^TA)^{-1}A^Tb)$ is projected (non-orthogonally) onto the feasible linear manifold $(Cx = d)$. Clearly this approach will be difficult if the unconstrained solution is not well-defined (ie. A^TA is semi-definite). This can happen even when the constrained solution is uniquely determined (ie. AZ is of full rank).

2.6.4 Infinite Weights

There is at least one other approach to this problem which we will mention briefly: the method of infinite weights. The idea is to solve an unconstrained least squares problem with infinite weights attached to the equality constraints. It can be argued that the usual procedure to reduce A to upper triangular form can be followed with one exception: when an infinite weight row is eliminating a component in an unweighted row then the Givens elimination is replaced with an elementary (Gauss) row elimination. See Bjorck and Duff [1980] for more details. An obvious advantage to this philosophy is that the unconstrained least squares algorithm is modified only slightly to handle constraints. On the other hand, in many applications the small number of equality constraints are rather dense which implies that the upper triangular factor of A will be dense also.

2.7 Exercises

1. Suppose that the matrix A can be partitioned

$$A = \begin{bmatrix} B & C \\ 0 & D \end{bmatrix}$$

where B is a square matrix of order p and D is a rectangular matrix with dimensions $(m-p)$ by $(n-p)$, and $m \geq n$. Prove that if A has rank n then B has rank p and D has rank $n-p$. Suppose that the intersection graph of $[B,C]$ is a complete graph: What conclusions would the George-Heath algorithm draw about the structure of the upper triangular form R, where $Q^T A = R$. Partition R as follows:

$$R = \begin{bmatrix} R_B & F \\ 0 & R_D \end{bmatrix}$$

Show that R_D^T is the Cholesky factor of $D^T D$. What is the apparent contradiction and how do you resolve it (algebraically) ? Suggest an improvement to the G-H method for structures of this kind.

2. Verify that the elimination method is a projection method (Section 2.6.1).

3. Consider problem (2.6.3) with $d = 0$ and $A^T A$ positive definite. Show that the solution can be written as

$$x = P_{G^{-1}}(A^T A)^{-1} A^T b$$

where $P_{G^{-1}}$ is a non-orthogonal projector (weighted by G^{-1}) that projects R^s onto the subspace $Cx = 0$.

4. Let A be m by n, C be t by n and Z be a matrix whose columns form a basis for $null(C)$ (assume that $t \leq n \leq m$). Prove that the following statements are equivalent:

   i. $rank(AZ) = n-t$,

   ii. $rank\begin{pmatrix} A \\ C \end{pmatrix} = n$,

   iii. $null(A) \cap null(C) = \{0\}$.

5. Consider the large constrained least squares problem (2.6.1). Let G represent the sequence of Gauss transformations (and row permutations) such that

$$GC^T = \begin{pmatrix} U \\ 0 \end{pmatrix}$$

where U is t by t and upper triangular. Show that a basis for $null(C)$ is given by

$$Z = G^T \begin{bmatrix} 0 \\ I_{n-t} \end{bmatrix}.$$

In practise G is not explicitly computed but is stored as a sequence of transformations (3 pieces of information per transformation). With this in mind, how can Z be used in an iterative projection method (Section 2.6.1)?

6. Consider the large constrained least squares problem (2.6.1). Let's assume that C is of full rank and t is small. Define $Z = \begin{bmatrix} -C_1^{-1} C_2 \\ I \end{bmatrix}$ where $C = (C_1, C_2)$ and C_1 is nonsingular. Assume that rows and columns of A have been permuted so that we can partition A:

$$\begin{bmatrix} 0 & A_{12} \\ A_{21} & A_{22} \end{bmatrix}$$

where A_{21} is p by t. Show that if $rank(A) = n$ then $p \geq t$. What is the structure of AZ ? Suggest an algorithm to find permutation matrices P and Q so that PAQ has the structure described above and p is small.

7. Consider the unconstrained large least squares problem where A has just a few dense rows. That is

$$A = \begin{pmatrix} A_1 \\ A_2 \end{pmatrix}$$

where A_2 consists of only a small number of dense rows and A_1 is sparse (with no dense rows). Consider the Peters-Wilkinson scheme and suggest the best way of handling A_2: You may assume that A is of full column rank however this does not imply that A_1 is of full column rank.

2.8 Research Problems

1. Thoroughly investigate the ideas outlined in question 6 and 7 above. Heath [1982] suggests a way of handling a few dense rows in the context of the George - Heath algorithm however it is necessary to assume that A_1 (defined in question 7) is of full column rank. Is there a modification to their method which does not need this assumption?

2.9 References

Bjorck A. and Duff I. [1980]. *A direct method for the solution of sparse linear least squares problems*, Linear Algebra and Its Applications, 43-67

George, A., Golub, G., Heath, M., and Plemmons, R. [1981]. *Least squares adjustment of large-scale geodetic networks by orthogonal decomposition*, Union Carbide Tech. Report ORNL/CSD-88, Oak Ridge.

George, A. and Heath, M. [1980]. *Solution of sparse linear least squares problems using Givens rotations*, Linear Algebra and Its Applications, 34, 69-83.

George and Ng [1983]. *On row and column orderings for sparse least squares problems*, SIAM Journal on Numerical Analysis 20, 326-344.

Heath, M. [1982]. *Some extensions of an algorithm for sparse linear least squares problems*, SISSC, 3, 223-237.

Lawson, C., and Hanson, R., [1974]. *Solving Least Squares Problems*, Prentice-Hall, Englewood-Cliffs, N.J.

Paige, C., Saunders, M. [1978]. *A bidiagonalization algorithm for sparse linear equations and least-squares problems*, SOL 78-19, Stanford University, Stanford Calif.

Paige, C., Saunders, M. [1982]. *LSQR: An algorithm for sparse linear equations and sparse least squares*, ACM Trans. on Math. Software, 8, 43-71.

Peters, G. and Wilkinson, J.H. [1970]. *The least squares problem and pseudo-inverses*, Comp. J. 13,309-316.

Weil, R. and Kettler, P. [1971]. *Rearranging matrices to block-angular form for decomposition and other algorithms*, Management Science 18, 98-108.

Chapter 3   Large Sparse Linear Programming

This chapter is written under the assumption that the reader is familiar with the fundamentals of linear programming. There are many texts concerned with linear programming - see Dantzig[1963], Orchard-Hays [1968], for example. Our purpose is to highlight some of the problems which are particular to the large sparse setting. We will not consider difficulties caused by degeneracy - indeed nondegeneracy will be assumed in this description.

3.1   The Problem and The Simplex Method

A linear programming problem can be written

$$minimize \ c^T x, \ Ax \geq b \tag{3.1.1}$$

where $x \in R^n$, A is m by n with $m \geq n$. The simplex method is usually the method of choice to solve (3.1.1). This famous algorithm is an example of a descent direction projection method for optimization. In general such a method generates a sequence of feasible points $\{x_k\}$. The next point, x_{k+1}, is obtained by correcting the current point: $x_{k+1} \leftarrow x_k + \alpha_k p_k$. The direction p_k is a projection of a gradient-related vector onto a facet of the feasible region. The steplength α_k is chosen to decrease the objective function and maintain feasibility.

The situation is simplified when all functions are linear. In particular, feasible descent directions and optimal steplengths are easy to compute. The simplex method begins at a feasible vertex and proceeds along a descent edge to an adjacent vertex. This process is repeated until optimality is reached (or an indication that the problem is unbounded or infeasible).

The number of steps required by the simplex method is unknown in advance and can be exponential (with respect to the problem size) in theory. In practise the number of steps is a linear function of the problem size and is usually considered quite acceptable.

Let x denote the current iterate, a feasible vertex, and let the active set which specifies this vertex be defined by $\overline{A}$: $\overline{A}$ consists of n rows of A and $\overline{A}x = \overline{b}$, where $\overline{b}$ is the corresponding right-hand-side vector. The Lagrange multipliers , λ, are defined by

$$\overline{A}^T \lambda = c \tag{3.1.2}$$

If $\lambda \geq 0$ then x is optimal otherwise there is a component $\lambda_s < 0$ and a descent edge can be realized if we move *away* from constraint s and stay *on* the remaining constraints in the active set. Clearly p is such a direction if

$$\overline{A}p = e_s \tag{3.1.3}$$

where e_s is a unit vector with component s equal to unity. The steplength will be defined by the first constraint which becomes active when moving away from x in the direction p. Notice that the new active set will differ from the old in exactly one row.

Gill and Murray[1973] proposed an implementation which is consistent with this viewpoint.

3.2 The Standard Form

It is much more common to express the linear programming problem in the standard form:

$$minimize \ c^T x, \ Ax = b, \ l \leq x \leq u \tag{3.2.1}$$

where A is m by n and $m < n$. The simplex method begins at a feasible vertex and always maintains the equalities $Ax = b$. In addition to the m equality constraints, a vertex will strictly satisfy $n-m$ of the simple inequality constraints. Therefore, a move from one vertex to an adjacent vertex will involve the release of one variable from its bound (upper or lower) and a free variable will reach its' bound (upper or lower).

Let x be a feasible vertex with x_N the vector of $n-m$ bounded, or *non-basic*, variables and x_B the vector of m free, or *basic*, variables. If we partition A accordingly, then

$$\begin{pmatrix} B & N \\ 0 & I_{n-m} \end{pmatrix} \begin{bmatrix} x_B \\ x_N \end{bmatrix} = \begin{bmatrix} b \\ d \end{bmatrix} \tag{3.2.2}$$

where d consists of the appropriate upper or lower bounds. The Lagrange multipliers, $\lambda^T = (\lambda_B^T, \lambda_N^T)$, are therefore defined by

$$B^T \lambda_B = c_B, \ \lambda_N = c_N - N^T \lambda_B$$

where (c_B^T, c_N^T) reflects a partition of c corresponding to (x_B^T, x_N^T).

The Lagrange multipliers λ_B are unrestricted in sign since they correspond to equality constraints: optimality or a descent direction is determined by the sign of λ_N. Suppose that x_s is at a lower bound and $\lambda_s < 0$. An appropriate descent direction is defined by

$$\begin{pmatrix} B & N \\ 0 & I_{n-m} \end{pmatrix} \begin{bmatrix} p_B \\ p_N \end{bmatrix} = e_s \tag{3.2.3}$$

or equivalently,

$$p_s = 1, \ Bp_B = -a_s,$$

and all other components of p are zero. (The vector a_s is column s of A.) The steplength α is then determined by the first variable to hit a bound as we move from x in the direction p. Clearly the new basis vector will differ form the old in exactly one column.

3.3 The Bartels - Golub [1969, 1971] Implementation

The standard form of the simplex method involves repeatedly solving systems of the form

$$Bx = y, \quad \text{and} \quad B^T w = v$$

where B is a square matrix of order m and consists of m columns of the matrix A. Each new basis matrix B is related to the old basis matrix in that they differ in exactly one column. Suppose that $B = LU$ and

$$B^+ = B + (a_r - a_s)e_i^T$$

where the ith column of B (which is currently column a_s of A) is replaced with column a_r of A. If we denote $L^{-1}a_r$ by $\bar{u}$ then

$$B^+ = LH$$

where $H = U + (\bar{u} - u_i)e_i^T$. If P is the permutation matrix which transfers column i to the end position and then shuffles columns $i+1,...$ to the left, then

$$B^+ P = L(HP),$$

where

$$HP = \begin{bmatrix} U_{11} & U_{12} \\ 0 & \bar{H} \end{bmatrix}$$

and U_{11} is of order $i-1$. $\bar{H}$ is clearly an upper Hessenberg matrix and can be stably reduced to upper triangular form by a sequence of Gauss transformations and row interchanges. (The work involved is $O(n^2)$). Each such transformation can be represented in three words and can be stored as a sequence (instead of explicitly updating L).

In summary, the Bartels-Golub implementation performs an initial LU factorization of the first basis matrix B; subsequent rank one modifications to B are processed by updating U in the manner described above and saving the Gauss transformations and interchanges as a sequence, G, say. If $G_k L^{-1} B_k = U_k$, where U_k is upper triangular, then the system $B_k x = y$ can be solved in the following way:

$$solve \ Lz = y, \ \bar{z} \leftarrow G_k z, \ solve \ U_k x = \bar{z}.$$

A system $B_k^T w = v$ can be solved

$$solve \ U_k^T z = v, \ \bar{w} \leftarrow G_k^T v, \ solve \ L^T w = \bar{w}.$$

Of course the initial lower triangular factor L (or L^{-1}) may be stored in factor form also, if desired. It may be necessary to re-factorize when the sequence G_k becomes too large, or when it is deemed wise to do so on stability grounds.

3.4 The Forrest-Tomlin[1972] Modification

The Bartels-Golub scheme is not appropriate, as it stands, for large sparse problems. One difficulty is that the matrix U must be accessed and modified when processing the rank-one update; fill will occur in arbitrary locations in U. Forrest and Tomlin suggested pivoting the ith row to the bottom:

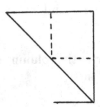

The last row can be 'zeroed' by successively applying Gauss transformations involving rows $(i,n),...(n-1,n)$. (Indeed only a subset of these row pairs need be applied since zeroes in row n can be exploited.) There will be no fill in U; however, the lack of pivoting indicates that the process is unstable. A compromise between the two approaches can be reached by using threshold pivoting: This will tend to limit fill, however a flexible data structure is still required.

3.5 Reducing The Bump (Reid[1982])

Suppose that we have just begun a basis change and $B^+ = LH$, where

$$H = U + (\bar{u} - u_i)e_i^T, \quad \bar{u} := L^{-1}a_r.$$

Pictorially, H has the form

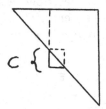

Consider the square 'bump' C outlined above. Reid suggested determining row and column permutations so that the size of the bump is reduced. Equivalently, let us determine the block upper triangular form of C (Section 1.2.1). The method of Tarjan [1972] can be used; however, advantage can be taken of the structure of C. Reid describeed a scheme that can be interpreted in the following way.

Assume that the (1,1) element of C is nonzero and construct the directed graph $G_D(C)$ (Section 1.2.1.2). Notice that every cycle in $G_D(G)$ must contain node v_1. It follows that G_C *has at most one strong component of cardinality greater than one* . All other strong components must have exactly one node each. These components (nodes) can be determined by

1. Construct node set N_1: successively locate and remove
   a node of indegree zero.
2. Construct node set N_3: successively locate and remove
   a node of outdegree zero.

The remaining node set, N_2, induces the one strong component of $G_D(C)$ of cardinality greater than one (it may not exist, of course).

The nodes can now be re-ordered: N_1 followed by N_2, followed by N_3. The ordering within sets N_1 and N_3 is defined by the order in which these sets were constructed; the order of the nodes within N_2 should be consistent with their original order in C - thus the spike structure persists in a smaller bump. This reduced bump can now be permuted into upper Hessenberg form. Finally, the new directed subgraph, defined by the upper Hessenberg submatrix, may not be strongly connected (if the original (1,1) entry of C was zero). In particular, step 1. above should be recursively applied to remove any single-node components.

3.6 Bumps and Spikes

Suppose that the matrix B has the following shape:

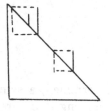

It is not difficult to see that if $B = LU$ then U has the following structure:

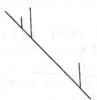

That is, fill occurs only in the spikes. (A *spike* is a column with at least one nonzero above the diagonal.) In addition, the only entries of L that will differ from those in B will occur in spike columns. A reasonable objective then is to order the columns and rows of B so that the number of spikes is minimized, subject to the diagonal being zero-free. This is surely an NP-hard problem although, to my knowledge, a proof is lacking. A more restrictive approach is to fix the diagonal with nonzeroes (ie. via matching - Section 1.2.1.1) and then minimize the number of spikes. This is equivalent to locating a minimum vertex feedback set in a directed graph: This problem is known to be NP-hard. (A vertex set is a feedback set if its removal from the graph leaves the graph acyclic.) Notice that if we partition a directed graph into its strong components and find the minimum feedback set in each component, then the full problem is solved also.

The usual approach is (roughly) this. Firstly, a zero-free diagonal is found via a matching algorithm. The algorithm of Tarjan[1972] is then applied to reduce B to block lower triangular form. (Recall that this block structure is essentially unique and is independent of the chosen diagonal.) Within each non-unitary block a heuristic is applied which symmetrically permutes columns and rows to minimize the number of spikes (vertex feedback set). Hellerman and Rarick [1972] suggested a heuristic to do the job. (Unfortunately, the number of spikes depends on the chosen diagonal.)

The real world contains roundoff error. Therefore, the good spike structure found above may have to be modified, to some degree, to allow for numerical pivoting. Saunders[1976] suggested a threshold pivoting compromise which limits interchanges to blocks (bumps).

3.7 The Method of Saunders [1976]

Once a good spike structure has been determined, a symmetric permutation can be applied to U, so that PUP^T has the following structure:

$$\begin{bmatrix} D & E \\ 0 & F \end{bmatrix}$$

where $\begin{bmatrix} E \\ F \end{bmatrix}$, is composed of only spike columns, in their original relative order, and D is diagonal.

The reason for this permutation is this. Now it is only necessary to maintain the matrix F in core when processing a new column. The matrices D and E can be stored in sparse format on secondary storage. When the ith column of U is replaced with column a_r the following steps can be performed.

1. Delete the ith column of U.
2. Add $L^{-1}a_r$ to the end of U.
3. Permute the ith row of U, v^T, to the bottom
4. Zero v^T by Gaussian elimination.

Notice that step 4 involves only matrix F since v^T has only zeroes to the left of column $m-p+1$, where F is order p. Whether or not F should be treated as a sparse or dense matrix is problem dependent.

3.8 Combining The Methods of Reid and Saunders (Gay [1979])

Gay suggested that Reid's bump reducing scheme can be combined with Saunders' approach. In particular, Reid's permutation strategy is applied to F, after a basis change. Gay reported significant sparsity gains on some very large problems.

3.9 Another Method Due to Saunders [1972]

Saunders described a method which uses an LQ decomposition of the basis matrix: $B = LQ$. A system $Bx = b$ can then be solved $LL^Ty = b$, $x \leftarrow B^Ty$. Paige has proven that the accuracy of the solution depends on $\chi(B)$, the condition number of B. (See Section 1.2.4, exercise 5, for more discussion.) A system $B^Tw = v$ can be solved $L^Tw = \bar{v} := Qv$. The vector $\bar{v}$ can then be updated from one iteration to the next - it is not necessary to store Q.

In order to be applicable to large sparse problems, it is necessary that B be permuted so that L is sparse. But

$$(P_1BP_2)(P_2^TB^TP_1^T) = P_1(BB^T)P_1^T$$

where P_1 and P_2 are permutation matrices: therefore, the column ordering of B plays no role in limiting fill. One possible approach to finding a good row ordering follows George and Heath (Section 1.2.4, exercise 5): determine the *structure* of BB^T and find a good permutation matrix such that the Cholesky factor of PBB^TP^T is sparse. Unfortunately, the basis will change by one column each iteration   it would not be efficient to find a new row ordering every step (followed by a full decomposition). In addition, in general we cannot expect L to remain sparse after several updates if we 'stick' with the original row ordering.

Saunders suggested that if the rows and columns of A are permuted to give the following profile for P_1AP_2:

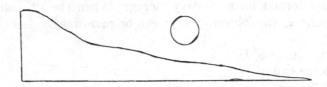

then we can expect, for an arbitrary basis B, that P_1BP_3 will be almost lower triangular:

Hence we can expect that the Cholesky factor of BB^T will have little fill since relatively few orthogonal transformations need be used. (Note that it is not necessary that we apply, or even determine, P_3: its *existence* is all that is needed.) Saunders suggests a heuristic algorithm, to be applied to A, which yields the form pictured above.

A major issue is how to process basis changes:

$$B^+ = B + (a_r - a_s)e_i^T$$

It follows that

$$B^+ B^{+T} = BB^T + a_r a_r^T - a_s a_s^T.$$

It might seem preferable to add column a_r before deleting a_s, in order to preserve intermediate positive definiteness; however, it is preferable to delete a column before adding a column, to avoid intermediate fill.

Deleting a Column

When column a_s is deleted, $\hat{B}$ will be deficient in rank, where $\hat{B} := B - a_s e_i^T$. Therefore, $\hat{B}$ does not have a Cholesky factor. Nevertheless, we can determine a lower triangular matrix $\hat{L}$ (singular) such that $\hat{L}\hat{L}^T = \hat{B}\hat{B}^T$.

Let $\bar{Q}$ be a sequence of Givens transformations such that

$$\begin{bmatrix} p^T & 0 \\ L & 0 \end{bmatrix} \tilde{Q} = \begin{bmatrix} o & \delta \\ \hat{L} & v \end{bmatrix}$$ (3.9.1)

where $Lp = a_s$, and $\tilde{Q}$ introduces zeroes in p^T from right to left. Notice that $L^{-1}a_s$ must be column i of Q: therefore, δ should equal 1. But taking (3.9.1) times its transpose yields

$$v = a_s, \text{ and } \hat{L}\hat{L}^T = LL^T - a_s a_s^T$$

which is the desired factorization of $\hat{B}\hat{B}^T$.

Adding a Column

After deleting a column we have the factorization

$$\hat{B}\hat{B}^T = \hat{L}\hat{L}^T$$

A new Cholesky factor, $\bar{L}$, is needed after adding column a_r:

$$\bar{L}\bar{L}^T = \bar{B}\bar{B}^T = \hat{L}\hat{L}^T + a_r a_r^T = (\hat{L}, a_r)\begin{bmatrix} \hat{L}^T \\ a_r^T \end{bmatrix}$$

Define $\bar{Q}$ to be a sequence of rotations such that

$$[\hat{L}, a_r]\bar{Q} = [\bar{L}, 0]$$

(This is similar to the row version of the QR decomposition described in Section 2.1.) The sequence of Givens rotations, $\bar{Q}$, may be much less than n, since sparsity in a_r can be exploited. Of course the nonzeroes of $\bar{L}$ will differ from the nonzeroes of L - a dynamic storage scheme is necessary.

In addition to this in-core method, Saunders proposed a product-form adaptation designed for using secondary storage efficiently. Neither Cholesky approach is popular however, relative to the methods described in Sections 3.6 and 3.7: the amount of fill tends to be greater than in the LU-based methods. Nevertheless, this general approach does have some appeal from a stability point of view and probably warrants further examination.

3.10 Steepest - Edge Simplex Algorithm (Goldfarb [1976])

Goldfarb showed that the steepest edge philosophy can be used in a large sparse setting. The steepest edge simplex method differs from the usual simplex method in one respect: in the former, the constraint which is 'dropped' yields the best *unit decrease* in the objective function whereas, in the latter, the constraint with the must negative multiplier (reduced cost) is dropped.

Let us consider the standard form of the simplex method and equations (3.2.2) and (3.2.3) in particular. It follows that the Lagrange multipliers corresponding to the nonbasic variables can be expressed as

$$\lambda_{N_i} = c^T p_i \tag{3.11.1}$$

where p_i is the direction in which we would move if variable x_{N_i} is released from its bound ($x^+ \leftarrow x + \alpha p_i$). The standard selection is $\lambda_{N_i} = \min\{\lambda_{N_i}\}$. (It is common however, to avoid computing all the multipliers, in large problems, and choose the first multiplier which is 'sufficiently' negative.) The steepest descent algorithm selects the most downhill direction:

$$\lambda_{N_i} = \min\{\frac{c^T p_i}{\|p_i\|}\}$$

The obvious apparent drawback is that the norm $\|p_i\|$ needs to be computed for every nonbasic variable, *in every iteration*. Goldfarb demonstrated that the norms can be efficiently recurred: the extra computation and storage is fairly minimal (computational experience of Goldfarb suggests an increase in work-per-step of about 25 percent) The hope is that these extra expenses will be offset by fewer iterations. Limited numerical results indicate that the method is (at least) competitive.

3.11 How Large is Large?

Classifying the size of a problem is machine dependent. The relative size of the linear programming problem to be solved should influence the choice of algorithm. A method which efficiently uses secondary storage is needed if the problem cannot be handled in primary memory. On the other hand, a secondary storage feature is unnecessary if the sparse techniques can be employed in main memory. To give the reader a 'feel' for the size of large linear programming problems solved in practise, we will briefly describe the classification scheme given by Bartels[1976].

Suppose that we are able to use $100k$ words of primary storage. An *LPP* is *small* if the matrix A is about 200 by 500 with 10% nonzeroes. In this case the matrix A should definitely be stored sparsely and it is probably worthwhile employing sparse factorization techniques. A *medium* problem is approximately 800 by 3000 with a density of about 1% It may be impossible to store both basis information and data in core at the same time; hence, the data should be kept on a secondary storage medium such as a disk. A *large* problem may have a matrix A with dimensions 8000 by 20,000 with .5% of the entries being nonzero. In this case both data and (some) basis information must be kept on a secondary storage devise. Finally, we may be faced with a *staggering* problem. In this case it is necessary that an underlying structure be found so that the problem becomes decomposable.

3.12 Exercises

1. State the steplength rules for the simplex algorithm described in Section 3.1 and 3.2.

2. Consider a linear programming problem expressed in the form (3.1.1) where the upper and lower bound constraints are included in $Ax \geq b$. Express this LPP in the form (3.2.1) and demonstrate that the algorithm described in Section 3.1 is equivalent to the algorithm described in Section 3.2. (The algorithm descriptions should be augmented with your steplength rules.)

3. Modify the simplex description in Section 3.2 to handle unbounded variables.

4. State the algorithm introduced in Section 3.5 in more detail: make reference to a particular data structure and determine the complexity.

5. Verify expression (3.11.1).

6. Let A be a nonsingular matrix with bipartite graph $G_B(A)$. Let M_1 and M_2 be 2 arbitrary perfect matchings in $G_B(A)$. Let their induced directed graphs be denoted by $G_D(A,M_1)$ and $G_D(A,M_2)$ respectively. Show that the strong component partitioning of the 2 directed graphs are equivalent.

   Let us try and minimize the number of spikes in the blocks defined by the strong components of $G_D(A,M_1)$ and $G_D(A,M_2)$ by restricting ourselves to *symmetric permutations*. Equivalently, we try and find a minimum vertex feedback set in each strong component. *Give an example* to illustrate that the minimum vertex feedback set in $G_D(A,M_1)$ may differ (dramatically ?) from the minimum vertex feedback set in $G_D(A,M_2)$.

7. The basic multipliers are determined by solving $B^T \lambda_B = c_B$. If the method of Section 3.9 is used, then this system can be reduced to

$$L^T \lambda_B = \bar{c}_B := Q c_B.$$

Explain how $\bar{c}_B$ can be updated, from one iteration to the next, so that the systems can be solved without storing Q.

3.13 Research Problems

1. Let A be an irreducible matrix (ie. $G_D(A)$ is strongly connected). Interpret the spike minimization problem in terms of the bipartite graph $G_B(A)$. What is the complexity of this problem ? Investigate heuristic approaches which are not dependent on the chosen diagonal.

2. The Cholesky-based method of Saunders (Section 3.9) attempts to maintain a

sparse Cholesky factor of BB^T. Suggest and investigate ways in which a *structured* Cholesky factor can be maintained.

3.14 References

Bartels R., and Golub G. [1969]. *The simplex method of linear programming using LU decomposition*, Comm. ACM 12, 266-268.

Bartels, R. [1971]. *A stabilization of the simplex method*, Numer. Math. 16, 414-434.

Bartels, R. [1976]. *Large Sparse Linear Programming*, Applied Matrix Methods Workshop, Johns Hopkins University, Baltimore, Maryland.

Dantzig, G. [1963]. *Linear Programming and Extensions*, Princeton University Press, Princeton, New Jersey.

Forrest J. and Tomlin [1972]. *Updating the triangular factors of the basis to maintain sparsity in the product form simplex method*, Mathematical Programming 2, 263-278.

Gill, P., and Murray, W., [1973]. *A numerically stable form of the simplex algorithm*, Algebra and Its Applics. 7, 99-138.

Gill, P., Murray, W., and Wright, M. [1981]. *Practical Optimization*, Academic Press, New York.

Gay, D.M [1979]. *On combining the schemes of Reid and Saunders for sparse LP bases*, in Sparse Matrix Proceedings 1978, Duff and Stewart (eds), SIAM, Philadelphia.

Goldfarb, D. [1976]. *Using the steepest-edge simplex algorithm to solve sparse linear programs*, in Sparse Matrix Computations, Bunch and Rose (eds), Academic Press, New York.

Hellerman, E., and Rarick, D. [1972]. *The partitioned pre-assigned pivot procedure*, in Sparse Matrices and Their Applications, Rose and Willoughby (eds), Plenum Press, New York.

Orchard-Hays, W. [1968]. *Advanced Linear-Programming Computing Techniques*, McGraw-Hill, New York

Reid, J. K. [1982]. *A sparsity-exploiting variant of the Bartels-Golub decomposition for linear programming bases*, Mathematical Programming 24, 55-69.

Saunders, M [1972]. *Large-scale linear programming using the Cholesky factorization*, PhD Dissertation, Stanford University, Stanford CA.

Saunders, M. [1976]. *A fast, stable implementation of the simplex method using Bartels-Golub updating*, in Sparse Matrix Computations, Bunch and Rose, eds.,Academic Press, New York.

Tarjan, R . [1972]. *Depth first search and linear graph algorithms*, SIAM J. Computing 1, 146-160.

In this chapter we are interested in the problem

$$minimize \ \|F(x)\|_2 \tag{4.1}$$

where $F: R^m \to R^n$, $m \geq n$, and F is at least continuously differentiable. In the case $m = n$ and $F(x_*) = 0$ then (4.1) can be expressed as the zero-finding problem:

$$solve \ F(x) = 0. \tag{4.2}$$

Techniques for (4.1) are often applicable to (4.2) however there are specialized methods for (4.2). As usual, in these notes we focus our attention on algorithmic considerations pertaining specifically to the large sparse situation.

4.1 The Jacobian Matrix

Efficient methods for (4.1) and (4.2) will involve the determination or estimation of the Jacobian matrix, $F'(x)$. Problems (4.1) and (4.2) are considered to be of the large sparse variety if $F'(x)$ is large and sparse. There are three common ways of determining (or approximating) $F'(x)$. Firstly, $F'(x)$ may be available analytically and hence it may be possible to differentiate $F(x)$ by hand (or possibly automatically) and write the code. This is a difficult, tedious and error-prone task - further development of automatic differentiation programs may offer some relief (eg. Speelpenning[1980]). A second possibility is to estimate $F'(x)$ by differencing: $F'(x)d \approx F(x+d) - F(x)$. This approach has particular appeal in the large sparse setting - we devote Section 4.1.1 to this topic. Finally, the Broyden update (or sparse Broyden update) can be used for the nonlinear equations problem. We discuss this in Section 4.2.2.

4.1.1 Estimation by Coloring

We will make the usual assumption that the component functions of $F(x)$ are highly interrelated and therefore it would be inefficient to evaluate each component separately. In other words, we will treat $F(x)$ as a black box which accepts the n-vector x and outputs the m-vector $F(x)$. (Note: In some cases a *small* number of equations are of a different variety and have dense gradient functions. In this situation it is important to divide F into 2 classes - each class is evaluated separately. Another possibility, which allows F to be treated as a black box and involves a partial Broyden update, is mentioned at the end of Section 4.2.2)

Let A be a matrix with the same sparsity pattern as $F'(x)$. Let $\{a_j : a_j \in C\}$ be a group of columns such that no two columns have a nonzero in the same row position. If d is a vector with components $d_j \neq 0$ if a_j belongs to C, and $d_j = 0$

otherwise, then

$$Ad = \sum_{j \epsilon C} d_j a_j,$$

and for each nonzero a_{ij}, $j \epsilon C$,

$$(Ad)_i = d_j a_{ij}.$$

Hence, if A is partitioned into p groups, $C_1, \ldots, C_p$ so that no two columns in the same group have a nonzero in the same row position, then A (or $F'(x)$) can be determined in p differences.

Curtis, Powell and Reid[1974] observed all of the above and suggested an algorithm to try and minimize p. Coleman and More [1982,1983], in an attempt to improve on the CPR scheme, interpreted the problem as a graph coloring problem. In particular, the intersection graph of A, $G_U(A)$, is constructed (implicitly): the nodes correspond to columns and two nodes are connected by an edge if the corresponding columns have a nonzero in the same row position. The problem of minimizing p is just this: color $G_U(A)$ in as few colors as possible. The set of columns belonging to one color form one group. (A coloring assigns different colors to adjacent nodes.) With this viewpoint, Coleman and More [1983] established that the problem is NP-hard and, in addition, developed improved heuristics for minimizing the number of groups.

The heuristics considered by Coleman and More are examples of *sequential coloring algorithms*. Let the vertices of $G_U(A)$ be ordered $v_1,...,v_n$. For $k=1,2,...,n$ the sequential algorithm assigns v_k the lowest possible color. Different orderings will result in different colorings. The method of Curtis, Powell and Reid is a sequential coloring algorithm where the ordering of the vertices is natural: vertex v_i corresponds to column a_i. Two particularily successful examples of ordering strategies are the smallest-last ordering of Matula, Marble, and Isaacson[1972] and the incidence degree ordering.

The smallest-last ordering is defined as follows. Assume that the vertices $v_{k+1},...,v_n$ have been selected, and choose v_k so that the degree of v_k in the subgraph induced by $V - \{v_{k+1}, \ldots, v_n\}$ is minimal. This ordering has a number of interesting properties such as the guarantee that the sequential coloring algorithm produces a coloring with at most

$$\max\{1 + \delta(G_0) : G_0 \ a \ subgraph \ of \ G\}$$

where $\delta(G_0)$ is the smallest degree of G_0. On the negative side, the sequential coloring algorithm with smallest-last ordering may produce arbitrarily bad results on bipartite graphs. The incidence degree ordering, on the other hand, is optimal for bipartite graphs. To define this ordering, assume that $v_1, \ldots, v_{k-1}$ have been selected and choose v_k so that the degree of v_k in the subgraph induced by $\{v_1, \ldots, v_k\}$ is maximal. Interestingly, the reverse of the ordering goes by the name *maximum cardinality search* (MCS) and is used to recognize graph chordality (Rose [1970], Tarjan [1975]). For further discussions concerning these orderings and their properties consult the paper by Coleman and More [1983].

If γ_{max} is the maximum number of nonzeroes in any row of A then A cannot be determined uniquely in fewer than γ_{max} differences by any method (this remark is *not* restricted to direct partition methods). Therefore, we can easily measure the success of any method: The coloring heuristics have proven to be *nearly optimal*, on 2 large sets of sparse 'test' matrices. That is, p is often equal to γ_{max}, and almost always satisfies $p \leq \gamma + 2$. Nevertheless, it is interesting to note that the methods are not optimal on some well-known matrix structures and, in addition, one can artificially construct examples where the colorings are arbitrarily far from optimal (Coleman and More' [1983]). Goldfarb and Toint[1983] have considered optimal schemes for specific matrix structures arising in a partial differential equations setting.

4.1.2 Indirect Estimation

The Jacobian estimation methods described above are *direct partitioning* methods. They are *direct* because once $F(x+d) - F(x)$ is determined, it is not necessary to do further calculations (except a single division) in order to solve for an element of the Jacobian matrix: we merely identify an element of $F'(x)$ with a component of $F(x+d)-F(x)$. The methods are *partition* methods because the columns of the Jacobian matrix belong to exactly one group.

It is possible to relax both qualifiers. A direct method which is not a partition of columns can be viewed as multi-coloring algorithm - a node is assigned several colors. I am not aware of any practical algorithms along these lines however Coleman and More' [1983] reported an example, due to Eisenstat, where an optimal multi-coloring betters the optimal (single) coloring.

An *optimal* differencing method can be defined if we demand neither direct determination nor a partition of columns. The method is optimal in that *exactly* γ_{max} differences will be needed. Newsam and Ramsdell[1981] proposed the following. Let A be the m by n sparse matrix of unknowns. Let D be an n by γ_{max} matrix of (differencing) vectors and let E be the m by γ_{max} matrix of differences (ie. column i of E is $F(x+d_i) - F(x)$), where d_i is column i of D.

The problem can now be restated: Is A uniquely determined by

$$minimize \; \{\|AD - E\|_F, \; A \in S(F')\} \tag{4.1.1}$$

where $S(F')$ is the space of all m by n matrices with the sparsity structure of F'. Conveniently, problem (4.1.1) is decomposable into m linear least squares problems,

$$minimize \; \|D^T_{[i]}a_{[i]} - E_i\|_2 \tag{4.1.2}$$

where $a_{[i]}$ corresponds to the γ_i nonzeroes of row i of A (to be determined), $D_{[i]}$ is the corresponding γ_i by γ_{max} submatrix of D and E_i^T is row i of E.

In order to achieve a unique solution it is necessary that each submatrix $D_{[i]}$ be of full rank γ_i. Notice that $D_{[i]}$ is just a selection of $\gamma_i \leq \gamma_{max}$ rows of D. One way to obtain the desired rank property is to choose the differencing directions so that every collection of γ_{max} or fewer rows of D is of full rank. Newsam and Ramsdell suggest using the Vandermonde matrix: row i is defined by $(1,a_i,...,a_i^{\gamma-1})$ where a_i, $i=1,...,m$ are distinct scalars. It is widely known that this matrix has the desired rank property. It is also widely known that the Vandermonde matrix has an undesired propery: it is very ill-conditioned.

Due to the conditioning problems mentioned above, the work involved to solve m least squares problems, and the practical success of the coloring approach, it does not appear that this method is a practical one. Nevertheless, it is quite interesting from a theoretical point of view; in addition, an efficient compromise between the two approaches might be useful for 'not-quite-sparse' problems. For example, a cross between the diagonal matrix approach (graph coloring) and the full matrix approach of this section might yield an efficient lower triangular matrix method.

4.2 Local Methods For Nonlinear Equations

In this section we discuss the (discrete) Newton and Broyden methods for solving $F(x) = 0$, in the large sparse setting. We will not discuss the algorithms of Brent[1973] and Brown[1966]: though interesting, it is not clear that they can be adapted to the large sparse situation. The question of how to enlarge the radius of convergence, by using a steplength procedure of some sort, will be discused in Section 4.4.

4.2.1 (Discrete) Newton's Method

Let A_k represent the Jacobian matrix, $F'(x_k)$ (or perhaps a finite difference approximation). Newton's method is simply

1. Solve $A_k s_k = -F(x_k)$
2. $x_{k+1} \leftarrow x_k + s_k$.

When Newton's method converges it usually does so rapidly - the rate of convergence is second order. The major sparsity considerations involve *obtaining* the matrix A_k (see Section 4.1) and *solving* the linear system.

The direct solution of the linear system can be achieved by methods discussed in chapters 1 and 2. We emphasize that if the QR or LQ decompositions are to be used, then it is not necessary to store Q. In addition, it can be assumed that the zero structure of A_k is invariant with k: hence, the storage allocation steps need be performed only in the initial iteration. (Note: If A_k is held 'constant' for several iterations, then the LQ decomposition has a distinct advantage - it is not necessary to retrieve Q (Exercise 5, Section 1.2.4).) If a sparse LU

decomposition is to be used, then it is not necessary to store L, unless A_k is held constant for several iterations. The sparsity of U will probably be greater than the sparsity of L (or R); however, pivoting for numerical purposes implies that a dynamic storage allocation scheme is necessary.

The linear system (step 1.) can be solved by an iterative method which typically will involve AA^T or A^TA (implicitly). An important observation, which we expand upon in the next section, is that it is possible to maintain a fast convergence rate without solving the linear system exactly. This observation is particularly relevant to iterative methods (for solving the linear system) because such a method can usually be stopped whenever the residual is deemed sufficiently small.

4.2.1.1 Inexact Newton Methods

It is reasonable to expect that the linear system

$$F'(x_k)s_k = -F(x_k) \qquad (4.2.1)$$

need not be solved exactly when far from the solution. Dembo, Eisenstat and Steihaug [1982] formalized this notion. Consider the following

Inexact Newton Algorithm:
1. Determine s_k satisfying $\|F'(x_k)s_k + F(x_k)\| \le \eta_k \|F(x_k)\|$
2. $x_{k+1} \leftarrow x_k + s_k$.

Questions of convergence can now be focused on the *forcing sequence* η_k. The following result establishes convergence.

Theorem 4.1 Assume $\eta_k \le \eta_{\max} < t < 1$. There exists $\epsilon > 0$ such that, if $\|x_o - x_*\| \le \epsilon$, then the sequence of inexact Newton iterates $\{x_k\}$ converges to x_*. Moreover, the convergence is linear:

$$\|x_{k+1} - x_*\|_* \le t\|x_k - x_*\|_*.$$

where $\|y\|_* := \|F'(x_*)y\|$.

This result is proven in Dembo, Eisenstat, and Steihaug [1982] and is stated under the assumptions that $F(x_*) = 0$, F is continuously differentiable in a neighbourhood of x_*, and $F'(x_*)$ is nonsingular.

Theoretically, it is usually considered important to establish that a method exhibits at least superlinear convergence. Dembo, Eisenstat and Steihaug proved that if the forcing sequence η_k converges to zero then the inexact Newton method is superlinearly convergent.

Theorem 4.2 Assume that the inexact Newton iterates $\{x_k\}$ converge to x_*. Then $x_k \to x_*$ superlinearly, if and only if,

$$\|F'(x_k)s_k + F(x_k)\| = o(\|F(x_k)\|).$$

Dennis and Moré [1974] proved a mathematically equivalent result which states that superlinear convergence is achieved, if and only if,

$$\|F'(x_k)s_k + F(x_k)\| = o(\|s_k\|).$$

The best way to define η_k is unknown: The only mathematical requirement is that $\eta_k \to 0$. Two popular choices are $\eta_k = k^{-1}$ and $\eta_k = \|F(x_k)\|$. (Indeed second-order convergence will result with the latter choice.)

The iterative methods described in Section 3 are all suitable for obtaining inexact solutions. Significant computational savings are to be expected when far from the solution - preliminary numerical experiments support this expectation. Nevertheless, crucial implementation problems remain. The (inner) convergence rate usually depends on $\chi^2(F')$: at some point it may be impossible to obtain the desired accuracy except possibly at a great expense - it may be wise to switch to a direct solver. At the very least, such a switch is awkward. Pre-conditioning the linear systems will help speed convergence: However, it is not clear how to best choose a pre-conditioner and how often to 'refresh' it.

4.2.2 Broyden's[1966] Method

Let B be an approximation to the Jacobian matrix $F'(x)$. If $x^+ (:= x - B^{-1}F(x))$ is the new iterate, then we would like to obtain a new improved Jacobian approximation, B^+, using the information $x, x^+, F(x), F(x^+)$ and B. If we define $s := x^+ - x$ and $y := F(x^+) - F(x)$, then $y = [\int_0^1 F'(x + \theta s)d\theta]s$. It is appropriate that B^+ resemble the average true Jacobian on $[x, x + s]$ by satisfying $B^+ s = y$. In addition, it is reasonable to demand that B^+ be 'near' to B. Therefore, the following optimization problem defines B^+:

$$\text{minimize } \{\|\bar{B} - B\|_F : \bar{B}s = y\} \tag{4.2.1}$$

Let C be the error matrix $\bar{B} - B$ and let r be the residual vector $y - Bs$. Problem (4.2.1) can be written as

$$\text{minimize } \{\frac{1}{2}\|C\|_F^2, \ Cs = r\} \tag{4.2.2}$$

But it is clear that (4.2.2) decomposes row-wise into n separate problems: For $j=1,...,n$

$$\text{minimize } \{\frac{1}{2}\|c_j\|_2^2, \ c_j^T s = r_j \} \tag{4.2.3}$$

A solution to (4.2.3) requires that $c_j = \lambda_j s$ for some λ_j. Substitution into the equation $c_j^T s = r_j$ yields $\lambda_j = \dfrac{r_j}{s^T s}$. Hence,

$$C = \lambda s^T, \quad \lambda_i = \frac{r_i}{s^T s}, \quad \text{for } i = 1, \dots n.$$

The Broyden update is therefore given by,

$$B^+ \leftarrow B + \frac{(y - Bs)s^T}{s^T s} \tag{4.2.4}$$

In summary, B^+ is the unique solution to

$$minimize \ \{\|\bar{B} - B\|_F, \ \bar{B} \in Q(y, s)\}$$

where $Q(y, s)$ is the affine set of matrices which send $s \to y$: B^+ is the projection of B onto $Q(y, s)$.

Broyden, Dennis and Moré [1973] established that Broyden's method is locally and superlinearly convergent, under reasonable assumptions. We note that this is achieved at great economy since 'extra' function evaluations are not needed to obtain an approximation to the Jacobian matrix.

Broyden's method can be implemented, using a QR factorization, so that each update can be stably processed in $O(n^2)$ operations. This method is not appropriate, as it stands, for large sparse problems, since updating will cause complete fill, in general.

Sparse Broyden (Schubert [1970])

A reasonable and practical request is that the sparsity of B reflect the sparsity of F'. Therefore, in addition to satisfying the quasi-Newton equation and being a minimal change from B, we demand that B^+ retain the sparsity of F'. In other words, if $C = B^+ - B$, then C is the solution to the problem

$$minimize \ \{\tfrac{1}{2} \|C\|_F^2, \ Cs = r, \ C \in S(F')\} \tag{4.2.5}$$

Again we are faced with a row-wise separable problem. Let S_j denote the space of n-vectors with the sparsity structure of row j of F'. That is, if $v \in S_j$ then $f'_{ji} = 0 \Longrightarrow v_i = 0$. If z is an n-vector then let $z_{[j]}$ denote the orthogonal projection of z onto S_j. For example, if $(f'_1)^T = (y, 0, 0, y, y, 0)$ and $z^T = (z_1, z_2, z_3, z_4, z_5, z_6)$ then $z_{[1]}^T = (z_1, 0, 0, z_4, z_5, 0)$. The jth row is defined by

$$minimize \ \{\tfrac{1}{2} \|c_j\|_2^2, \ c_j^T s_{[j]} = r_j\}$$

Therefore,

$$c_j = \lambda_j s_{[j]} \text{ and } \lambda_j = \frac{r_j}{s_{[j]}^T s_{[j]}}, \quad provided \ s_{[j]} \neq 0.$$

Define $\lambda_j = 0$ if $s_{[j]} = 0$. Note that if $s_{[j]} = 0$ then $y_j = 0$ and hence $r_j = 0$. To see this consider that for some $\theta \in [0,1]$,

$$y_j = [(F'(x + \theta s))s]_j = [f_j'(x + \theta s)]_{[j]}^T s_{[j]} = 0.$$

The correction matrix C can be written as the sum of n rank-one modifications:

$$C = \sum_1^n \lambda_j e_j s_{[j]}^T$$

In summary, B^+ is the unique solution to the optimization problem

$$minimize \ \{\|\bar{B} - B\|_F, \ \bar{B} \in Q(y,s) \bigcap S(F')\}$$

where $Q(y,s)$ is the affine set of matrices which send $s \to y$.

Marwil[1979] has proven that the sparse Broyden update is locally and super-linearly convergent. Unfortunately, the fact that we are dealing with a rank-n modification suggests that there is no way to efficiently update a factorization of B. (It is possible to implement the sparse Broyden update using p rank-one updates where p is the number of rows with *distinct* sparsity patterns. This suggests a potential approach which allows some 'fill' in order to decrease p.)

It is not clear that the sparse Broyden update represents a practical improvement over a finite-difference estimate. A full factorization is needed in either case; a sparse finite-difference approximation can usually be obtained using a small number of differences and yields a more accurate approximation. However, a hybrid algorithm might be useful when F' consists of a few dense rows. That is, suppose that F' can be partitioned $F' = \begin{bmatrix} J_1 \\ J_2 \end{bmatrix}$, where J_1 is large and sparse and J_2 has a few dense rows. Then J_1 can be estimated by finite differences and J_2 by a Broyden update. If a graph coloring approach is used to estimate J_1 then it is only necessary to color the intersection graph $G_U(J_1)$. Note that this approach can be implemented while still treating $F(x)$ as a black box.

4.2.3 Updating The Factors (Dennis and Marwil [1981])

A major difficulty with the sparse Broyden update is that there is no efficient method to update a factorization of B: A sparse factorization of B is required in every iteration. In order to circumvent this problem, Dennis and Marwil suggested that a 'quasi-Newton condition' be imposed on an LU factorization *directly*. That is, if $PB = LU$ then we can implicitly define the new improved matrix B^+ by imposing a sparse Broyden update on U to obtain U^+. If $L^+ = L$ and $P^+ = P$ then B^+ is implicitly defined by $P^+ B^+ = L^+ U^+$.

More specifically, if $B^+ s = y$ then we can define vector v such that $L^+ v = P^+ y$, and $U^+ s = v$; if we let $L^+ = L$ and $P^+ = P$ then it is reasonable to define U^+ by the optimization problem:

$$minimize \ \{\|\bar{U} - U\|_F: \ \bar{U} s = v, \ \bar{U} \in S(U)\}$$

where $S(U)$ is the space of matrices with the sparsity structure of U. Clearly the sparse Broyden updating technique can be applied to U. It is important to note that the case $s_{[j]} = 0$ is troublesome here since it does not follow that $v_j = 0$. Hence the intersection $S(U) \cap Q(v,s)$ may be empty. In this case, the sparse Broyden update, which leaves row j of U unchanged, solves the optimization problem

$$minimize\ \{\|\bar{U} - U\|: \bar{U} \in S(U),\ \bar{U}\ is\ a\ nearest\ point\ Q(v,s)\}$$

There are many other details - which we will not discuss in these notes - which are important to a practical realization of these ideas. For example, a periodic restart is necessary - see Dennis and Marwil[1981] for more details. Indeed it may be best to view this scheme as an alternative to holding the Jacobian approximation constant (or unchanged) for a 'fixed' number of iterations.

4.3 Local Methods For Nonlinear Least Squares

Let us assume that $f_i(x)$ is twice continuously differentiable for $i=1,...,m$. It is natural to consider solving the nonlinear least squares problem, $minimize\ f(x) := \frac{1}{2}\|F(x)\|_2^2$, by Newton's method. But the gradient of f is

$$\nabla f(x) = F'(x)^T F(x) \tag{4.3.1}$$

and the Hessian of f can be expressed as

$$\nabla^2 f(x) = [F'(x)^T F'(x) + \sum_1^m f_i(x)\nabla^2 f_i(x)] \tag{4.3.2}$$

Therefore Newton's method reduces to

1. Solve $[F'(x)^T F'(x) + \sum_1^m f_i(x)\nabla^2 f_i(x)]s = -\sum_1^m f_i(x)\nabla f_i(x)$
2. $x^+ \leftarrow x + s$

Observe: if $F(x_*) = 0$, then the Hessian of f, evaluated at x_*, is just $F'(x_*)^T F'(x_*)$. Moreover, if the residuals at the solution, $f_i(x_*)$, are of 'small' magnitude or if the functions $f_i(x)$ are 'almost' linear then the Hessian of f, evaluated at x_*, is 'almost' $F'(x_*)^T F'(x_*)$. This observaton divides the approaches to the nonlinear least squares problem into two camps: those that ignore the term $\sum f_i(x)\nabla^2 f_i(x)$ and those that do not. The latter strategy is further divided into those methods which attempt to exploit the structure of the Hessian matrix (4.3.2) and those that ignore the structure and treat the minimization of $f(x)$ as an *ordinary* nonlinear optimization problem. In these notes we will consider only the first approach, which ignores the term $\sum f_i(x)\nabla^2 f_i(x)$, and is particularily suitable for low residual and 'almost' linear problems. Dennis, Gay and Welsch[1983] and Gill and Murray[1978] have proposed algorithms which exploit the structure of (4.3.2) while maintaining an approximation to the Hessian of f which includes the term $\sum f_i(x)\nabla^2 f_i(x)$. Nazareth [1980] discusses these methods and proposes an interesting hybrid algorithm.

4.3.1 The Gauss-Newton Method

The (local) Gauss-Newton method involves a linearization of the functions and can be stated as follows:

1. Minimize $\|F(x) + F'(x)p\|$
2. $x^+ \leftarrow x + p$

If $F(x_*) = 0$ then the Gauss-Newton method converges locally at a second order rate. If the residuals are 'small' or the functions $f_i(x)$, $i = 1,...,m$ are almost linear, then the method is locally convergent with a 'fast' linear convergence rate.

An implementation designed for solving large sparse problems, based on solving the linear least squares problem directly, may involve any of the sparse linear least squares approaches discussed in Chapter 2. An inexact approach is also possible (see Section 4.2.1.1) using one of the indirect linear solvers, discussed in Section 1.3, on the normal equations.

4.4 Global Methods

In this section we will discuss mechanisms for enlarging the radius of convergence of the local methods discussed in Sections 4.2-4.4. In particular, we will describe two 'trust region' procedures which are currently implemented in MINPACK-1 (More, Garbow, Hillstrom [1980]): implementations of the Levenberg-Marquardt[1944,1963] algorithm and Powell's 'dogleg' strategy [1970], will be discussed. Both algorithms are designed for dense problems - we will suggest possible modifications to handle sparsity in Sections 4.4.2 and 4.4.5. Both algorithms are applicable to the nonlinear least squares problem as well as nonlinear equations. However, the L-M method is usually used for nonlinear least squares and the 'dogleg' strategy is commonly reserved for the square case.

The algorithms have strong convergence properties: in Chapter 5, a general discussion about these properties will be presented.

4.4.1 The Levenberg-Marquardt Algorithm (More [1978])

This algorithm is designed to increase the radius of convergence of the Gauss-Newton iteration. In particular, we motivate and describe an algorithm that satisfies

$$\lim_{k \to \infty} \|[F'(z_k)]^T F(z_k)\| = 0$$

under fairly weak assumptions.

The Locally Constrained Problem

The unconstrained Gauss-Newton step, p_u, is a minimizer of the function $\Psi(p) = \|F(x) + F'(x)p\|$. Since $F(p) \ (= F(x+p))$ is *not* a linear function, a linear approximation will usually not be an accurate representation of $F(p)$ outside a neighbourhood of x. Hence it is reasonable to compute a correction to x by solving the *constrained* problem

$$\min\{\Psi(p): \|p\| \leq \Delta\} \tag{4.4.1}$$

where Δ represents a radius in which $\Psi^2(p)$ is to be 'trusted' to accurately reflect $f(p)$. (Note: *The constraint should really be* $\|Dp\| \leq \Delta$, *where D is a diagonal scaling matrix.* However, we will drop the scaling matrix in this brief description, for simplicity.) The choice of Δ is important, of course, and we will discuss this shortly.

In our development to follow it is more convenient to replace (4.4.1) with the equivalent problem

$$\min \{ \Psi^2(p): \|p\|^2 \leq \Delta^2\} \tag{4.4.2}$$

Notice that Ψ^2 is a *convex* function and is *strictly convex* if and only if the Jacobian matrix, $J := F'(x)$, is of full rank. The constraint is also convex and therefore every local minimizer of (4.4.2) is a global minimizer. Clearly the *minimum norm global* unconstrained minimizer of Ψ^2, p_u, solves (4.4.2) if and only if $\|p_u\|^2 \leq \Delta^2$. Otherwise, $\|p_u\|^2 > \delta^2$ and the solution to (4.4.2), p_*, satisfies $\|p_*\|^2 = \Delta^2$; there exists $\lambda_* > 0$ such that

$$(J^T Jp + J^T F) = -\lambda_* p$$

or, equivalently

$$(J^T J + \lambda_* I)p = -J^T F \tag{4.4.3}$$

Notice that the matrix $J^T J + \lambda I$ is positive definite for all $\lambda > 0$. The remarks above suggest the following scheme to solve (4.4.2):

Algorithm To Solve The Locally Constrained Problem
  1. Determine p_u: solve $\min\|Jp - F\|$.
     (p_u is the LS solution of minimum norm.)
  2. If $\|p_u\| \leq \Delta$ then $p_* \leftarrow p_u$.
  3. If $\|p_u\| > \Delta$, determine λ_*, p_* such that
     $(J^T J + \lambda_* I)p_* = -J^T F$, $\|p_*\| = \Delta$, $\lambda_* > 0$.

A viable approach to step 3. is to treat $p = -(J^T J + \lambda I)^{-1}J^T F$ as a function of λ: $p = p(\lambda)$. The nonlinear equation $\|p(\lambda)\| = \Delta$ can be solved using an iterative technique. Indeed, $\phi(\lambda) := \|p(\lambda)\| - \Delta$ is a continuous, strictly decreasing function on $[0, \infty)$, and approaches $-\Delta$ at infinity. (Note that , even in the rank deficient case, $p(\lambda)$ approaches p_u as $\lambda \to 0$.) Hence there is a unique λ_* such that $\phi(\lambda_*) = 0$. Moré [1978] describes a scheme, based on Hebden[1973], which generates a convergent sequence $\lambda_0, \lambda_1, \cdots$. In practise convergence occurs rapidly; less than 2 iterations are needed, on average, to obtain suitable accuracy.

Each evaluation of $\phi(\lambda)$ will involve solving

$$(J^T J + \lambda I)p = -J^T F, \lambda > 0 \tag{4.4.4}$$

and we describe that process next.

Solving The Structured Linear Least Squares Problem
    The solution of the positive definite system (4.4.4) is also the least squares solution to

$$\begin{Bmatrix} J \\ \lambda^{\frac{1}{2}} I \end{Bmatrix} p \approx -\begin{pmatrix} F \\ 0 \end{pmatrix} \tag{4.4.5}$$

System (4.4.5) can be solved by the QR decomposition. However, the algorithm above initially requires a least squares solution solution to

$$Jp \approx -F \tag{4.4.6}$$

and J may be rank deficient: we first compute the QR decomposition of J with column pivoting

$$QJ\pi = \begin{bmatrix} T & S \\ 0 & 0 \end{bmatrix} \tag{4.4.7}$$

where T is a nonsingular upper triangular matrix of $rank(J)$ order. A least squares solution to (4.4.6) is given by

$$\tilde{p}_* = -\pi \begin{bmatrix} T^{-1} & 0 \\ 0 & 0 \end{bmatrix} QF \tag{4.4.8}$$

Note that $\tilde{p}_*$ is *not* the least squares solution of minimum norm (Section 4.4.3). To solve (4.4.5), first observe that

$$\begin{bmatrix} Q & 0 \\ 0 & \pi^T \end{bmatrix} \begin{bmatrix} J \\ \lambda^{\frac12} I \end{bmatrix} \pi = \begin{bmatrix} R \\ 0 \\ \lambda^{\frac12} I \end{bmatrix} \tag{4.4.9}$$

where $R = (T,S)$. (R is singular if and only if S exists.) The matrix $(R^T, 0, \lambda^{\frac12} I)^T$ can now be reduced to upper triangular form with a sequence, W, of $\dfrac{n(n+1)}{2}$ Givens rotations,

$$W \begin{bmatrix} R \\ 0 \\ \lambda^{\frac12} I \end{bmatrix} = \begin{bmatrix} R_\lambda \\ 0 \end{bmatrix} \tag{4.4.10}$$

yielding the solution $p = -\pi R_\lambda^{-1} u$ where $W\binom{QF}{0} = \binom{u}{v}$. It is important to note that only this last step need be repeated if λ is changed (MINPACK-1 saves a copy of the matrix R).

The Trust Region Size

The choice of Δ should reflect the degree to which we trust the linear approximation to F. A particular measure is given by the ratio

$$\rho(p) = \frac{\|F(x)\|^2 - \|F(x + p)\|^2}{\|F(x)\|^2 - \|F(x) + J(x)p\|^2} \tag{4.4.11}$$

where p is the computed solution to the constrained problem (4.4.2). That is, we are measuring the ratio of the actual decrease to the predicted decrease. If $\rho(p)$ is close to unity (eg. $\rho(p) \geq 3/4$) then we may want to *increase* Δ; if $\rho(p)$ is close to zero, (eg. $\rho(p) \leq 1/4$) then we may want to decrease Δ. More' gives specific rules which allow for statement 4 in the algorithm to follow. Finally, if $x + p$ does not yield a sufficient decrease (eg. $\rho(p) \leq .0001$) then p should not be accepted and Δ should be reduced.

The Algorithm

1. Let $\sigma \in (0,1)$. Determine p_*.

   If $\|p_*\| \leq (1+\sigma)\delta_k$ then set $\lambda_k = 0$.

   Otherwise, determine $\lambda_k > 0$ such that if

$$\begin{bmatrix} J_k \\ \lambda_k^{\frac12} I \end{bmatrix} p_k = -\begin{bmatrix} F_k \\ 0 \end{bmatrix} , \text{ in the least squares sense,}$$

then

$$(1-\sigma)\delta_k \leq \|p_k\| \leq (1+\sigma)\delta_k$$

2. Compute the ratio ρ_k of actual to predicted reduction.
3. If $\rho_k \leq .0001$, set $x_{k+1} \leftarrow x_k$, $J_{k+1} \leftarrow J_k$.
   If $\rho_k > .0001$, set $x_{k+1} \leftarrow x_k + p_k$, compute J_{k+1}.
4. If $\rho_k \leq \frac{1}{4}$, set $\Delta_{k+1} \epsilon [\frac{1}{10}\Delta_k, \frac{1}{2}\Delta_k]$.

   If $\rho_k \epsilon [\frac{1}{4}, \frac{3}{4}]$ and $\lambda_k = 0$,

   or $\rho_k \geq \frac{3}{4}$ then set $\Delta_{k+1} \leftarrow 2\|p_k\|$.

4.4.2 The L-M Algorithm and Sparse Considerations

In this section we will briefly discuss the possibility of extending the L-M algorithm to the large sparse setting. In particular, we consider adapting a sparse QR decomposition to the solution of (4.4.2). It is also possible to use other sparse decompositions and direct techniques such as the Peters-Wilkinson scheme mentioned in Section 2.4. A sparse inexact iterative approach is also possible: we postpone a discussion of this type of method until Chapter 5.

The sparse QR decomposition determines a column ordering of J such that the Cholesky factor of $J^T J$ is sparse (Section 1.2.4, question 5). Clearly this is incompatible with the determination of π (4.4.7). Therefore it is necessary that the approach to the rank deficient case be changed. (Heath [1982] suggested several ideas on this topic.) There is no need to store Q: the Givens rotations can be applied and then discarded.

Several least squares systems of the form

$$\begin{pmatrix} J \\ \lambda^{\frac{1}{2}} I \end{pmatrix} p \approx -\begin{pmatrix} F \\ 0 \end{pmatrix}$$

must be solved for different values of λ. The procedure described in Section 4.4.1 involves computing the QR decomposition of J, saving a copy of R, and then obtaining R_λ, from R, via a sequence of Givens transformations W. (Note that the structure of R equals the structure of R_λ.) In the large sparse setting it may appear that it is more space-efficient to avoid storing a copy of R and to retrieve R via W: This is incorrect - it is always preferable to save a copy of R. If space constraints do not allow this, then it is necessary to compute the factorization from scratch.

4.4.3 Notes On The MINPACK-1 Software

The description given in the previous section is consistent with the paper of More [1978] however the software package, MINPACK-1, differs slightly. In particular, there is a potential problem which arises with the use of $\tilde{p}_u$, instead of p_u, in the rank-deficient case. If $\|\tilde{p}_u\| > \Delta$ and $\|p_u\| < \Delta$, then the equation $\|p(\lambda)\| = \Delta$, $\lambda > 0$, has no solution since $\|p(0)\| = \|p_u\| < \Delta$, and the function $\|p(\lambda)\| - \Delta$ decreases as λ increases. MINPACK-1 avoids this trap: a solution to $\|p(\lambda)\| = \Delta$ is not demanded if $\|\tilde{p}_u\| > \Delta$. Instead, the relationship $p(\lambda) \to p_u$ as $\lambda \to 0$ is used to determine a solution p_* which is 'close' to p_u.

A few other remarks on the rank deficient case are appropriate: A procedure to reliably solve the rank deficient problem involves the determination of rank. This is a delicate numerical problem which is most stably solved by the singular value decomposition. The procedure described in Section 4.4.2 will, in exact arithmetic, determine the rank of J: $rank(J) = |T|$. In finite precision, the rank of J may differ. MINPACK-1 makes no attempt to guess the correct rank: R is partitioned $R = (T,S)$ to avoid (absolute) zero diagonal elements. Hence we can conclude that $rank(J) \le |T|$. The matrix T may contain very small diagonal elements, possibly reflecting rank deficiency, which may result in a very large computed vector $\tilde{p}_u$. Nevertheless, the computed solution to the sub-problem, p_*, is restricted in size by Δ and *in this sense the algorithm is not sensitive to rank deficiency problems*. Of course, if the Jacobian is rank deficient at the solution, x_*, then convergence may be very slow.

4.4.4 Powell's [1970] Hybrid Algorithm in MINPACK-1

This trust region algorithm is designed to increase the radius of convergence of Newton-like methods for systems of nonlinear equations. This method is applicable to nonlinear least squares also, however we will restrict our discussion to the square case. A Levenberg-Marquardt algorithm is probably more appropriate for nonlinear least squares since accurate information is usually demanded for the over-determined problem and the L-M algorithm takes better advantage of this exact information. A nonlinear equation solver may use approximate Jacobian information, such as the Broyden update, and therefore the more expensive L-M algorithm is probably not justified.

In order to measure 'progress', a mechanism is needed for determining when a point $x + s$ is 'better' than a point x. It is convenient to use the 2-norm: the point $x + s$ is 'better' if $\|F(x + s)\|_2 < \|F(x)\|_2$. Note that this measure is not foolproof since it easy to construct examples where x_* minimizes $\|F(x)\|_2$ but $F(x_*) \ne 0$ and there exists a vector $\overline{x}$ such that $F(\overline{x}) = 0$. Nevertheless, the least squares measure is often quite satisfactory in practise.

The approach we will take is to minimize $f(x) := \frac{1}{2} \|F(x)\|_2^2$, under the expectation that $F(x_*) = 0$. If we let B denote an approximation to the Jacobian matrix of F then the gradient of f is approximately $g(x) := B^T F(x)$, which we will call the *gradient direction*. The Hessian of f is approximated by $B^T B$. As in the L-M algorithm, we will approximate $f(x+s)$ on the ball $\|s\|_2 \leq \Delta$ by a quadratic

$$\Psi^2(s) = \tfrac{1}{2}\ s^T B^T B s + (B^T F)^T s$$

(Again, the constraint should really be $\|Ds\|_2 \leq \Delta$ where D is a diagonal scaling matrix, however we will omit D in this description, for simplicity.)

The Newton correction, $B s_N = -F$, is highly desirable since then a superlinear convergence rate is usually achieved. However, s_N is not always a suitable correction since it is possible that $f(x+s_N) > f(x)$. On the other hand, the negative gradient direction, $s_G := -B^T F$, is attractive for ensuring convergence. In particular, if we define a *Cauchy point*, s_C, by

$$\Psi(s_C) = \min\{\Psi(s): s = \alpha s_G,\ \|s\| \leq \Delta\} \tag{4.4.12}$$

then fairly strong convergence results can be obtained provided

$$\Psi(s) \leq \beta \Psi(s_C),\ \beta > 0 \tag{4.4.13}$$

That is, the value of $\Psi(s)$ is at least as good as a constant factor times the Cauchy value. (Note that if $\|s_C\| \leq \Delta$ then $\alpha = [\dfrac{-g^T s_G}{\|B s_G\|^2}]$.)

The Locally Constrained Problem

An approximate solution to problem (4.4.2) can be obtained if s is further restricted. For example, if s is constrained to satisfy $s = \alpha s_G$, then problem (4.4.12) results; the solution s_C can be viewed as an approximate solution to (4.4.2). Other possibilities include minimizing $\Psi(s)$ over a 'slice' of the ball $\|s\| \leq \Delta$, and minimizing $\Psi(s)$ over a path. Powell[1970] chooses a hybrid path, P_H, defined by the line segment joining the origin to the optimum point along s_G, s_G^*, followed by the line segment joining s_G^* to the Newton point, s_N. (We will assume that B is nonsingular: for further remarks see Section 4.4.6.) Pictorially, we have the following situation:

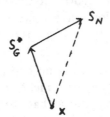

The correction to x, s_*, is defined by the optimization problem:

$$\min\ \{\Psi^2(s): s \in P_H,\ \|s\| \leq \Delta\} \tag{4.4.14}$$

Since Ψ^2 is a convex function, Ψ^2 is convex on the lines $(0, s_G^*)$ and (s_G^*, s_N).

Therefore, s_G^* minimizes Ψ^2 over $\{s: s = \theta s_G^*; 0 \leq \theta \leq 1\}$ and s_N minimizes Ψ^2 over $\{s: s = s_G^* + \theta(s_N - s_G^*); 0 \leq \theta \leq 1\}$. Therefore Ψ^2 is strictly decreasing along P_H. In addition, it is easy to show that $-g^T(s_N - s_G^*) > 0$ provided B is of full rank (hence $\|s\|$ is strictly increasing as s follows P_H.) Consequently, we are led to the following three possibilities:

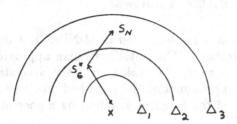

i) $\|s_N\| \leq \Delta: s_* \leftarrow s_N$
ii) $\|s_N\| > \Delta$, $\|s_G^*\| \leq \Delta$: $s_* \leftarrow s_G^* + \gamma(s_N - s_G^*)$, $\|s_*\| = \Delta$,
iii) $\|s_N\| > \Delta$, $\|s_G^*\| > \Delta$: $s_* \leftarrow \alpha s_G^*$, $\|s_*\| = \Delta$.

These three situations, in their given order, essentially define the algorithm to solve (4.4.14). Note that exactly one linear system need be solved. The parameters α, γ are in the interval [0,1] and are easy to compute.

The solution s_* cleverly interpolates between a scaled gradient direction and the Newton step. If the Newton step is unsuitable, (ie. $\|s_N\|$ is too large) then the correction p_* tends toward the steepest descent direction. As $\|s_N\|$ decreases, relative to Δ, the correction tends toward the Newton step.

The Trust Region Size

The trust region size, Δ, is monitored precisely as described in the Levenberg-Marquardt method (Section 4.4.1). The final algorithm given at the end of Section 4.4.1 describes this hybrid algorithm also, provided step 1. is replaced with i, ii, and iii above.

4.4.5 The Dogleg Algorithm and Sparse Considerations

A sparse direct method can be developed using a sparse QR or LU decomposition. The comments and observations concerning sparse updates and sparse finite differences mentioned in Sections 4.1 and 4.2 are applicable here and we will not repeat them. However, it was noted that it may be most efficient to obtain an 'exact' Jacobian approximation (by sparse finite differences) instead of using a sparse Broyden update. If this is the case, then a Levenberg-Marquardt algorithm may be preferable to this hybrid algorithm since the L-M algorithm should take greater advantage of exact information. On the other hand, it is possible to re-organize steps i,ii, and iii above so that when $s_* \leftarrow \alpha s_G$ then s_N *is not computed.* (Note that $\|s_G^*\| > \Delta => \|s_N\| > \Delta$.) This could result in considerable computational savings.

A discussion concerning an inexact iterative approach to this type of method will be given in Chapter 5.

4.4.6 Notes on The MINPACK-1 Software

Powell's hybrid algorithm is implemented in MINPACK-1 using the Broyden update and the QR factorization. The initial Jacobian approximate is obtained using finite differences, (or is explicitly evaluated if the derivative functions are supplied). The Jacobian approximation is refreshed via finite differences (or evaluation) when Δ is reduced too frequently, reflecting a poor quadratic approximation.

If R has a very small diagonal element then B is probably close to singularity and s_N will probably be very large. In this case the computed s_* will lie close to the gradient direction; the size of s_* will be bounded by Δ. If R has an *absolute* zero diagonal element, then MINPACK-1 perturbs this element by a small amount (relative to the size of the largest diagonal element). Again the result will be that s_N will be very large: s_* will lie in the gradient direction (essentially) with a magnitude bounded by Δ. *In this sense the implementation is insensitive to difficulties caused by singularity.* Of course, if the Jacobian is singular at the solution, then convergence may be very slow since the method reduces to a steepest descent algorithm.

4.4.7 Exercises

1. Prove that if γ_{max} is the maximum number of nonzeroes in any row of A and A is uniquely determined by $Ad_1, Ad_2, \ldots, Ad_p$, then $p \geq \gamma_{max}$.

2. A sequential coloring scheme orders the nodes of a graph, $x_1, \ldots, x_n$ and then executes

   $i \leftarrow 1$
   while $i \leq n$ do
       assign to node x_i the lowest possible color
   od

   Give an example where 2 different orderings result in a *dramatic* difference in the number of colors used.

3. Under suitable assumptions, prove that the superlinearity characterization in Theorem 4.2 is equivalent to the Dennis and Moré [1974] condition

   $$\|F'(x_k)s_k + F(x_k)\| = o(\|s_k\|).$$

4. Show that it is possible to implement the *sparse* Broyden update using p rank-1 updates where p is the number of rows with distinct sparsity patterns.

5. With regard to the comments in Section 4.4.2, show that the space required to store R is always less than that required to store W.

6. Prove that $p(\lambda) \to p_*$ as $\lambda \to 0^+$ where $p(\lambda) = -(J^T J + \lambda I)^{-1} J^T F$ and p_* is the least squares solution of minimum norm to $Jp \approx -F$.

7. Let P_H be the dogleg path defined in Section 4.4.4. Prove that $\|s\|$ is strictly increasing as s follows P_H. Prove that $\|s_G^*\| > \Delta \implies \|s_N\| > \Delta$.

8. Suggest Broyden and sparse Broyden updates for the nonlinear least squares problem. Discuss possible reasons for the lack of success of Broyden updates for the nonlinear least squares problem.

4.4.8 Research Problems

1. The sparse Broyden update can be implemented using p rank-1 updates where p is the number of rows with *distinct* sparsity patterns. Investigate an approach which allows some 'fill' in B in order to decrease p.

2. Investigate a trust region approach for the nonlinear least squares problem which is based on $\|\cdot\|_\infty$ or $\|\cdot\|_1$.

4.4.9 References

Brent, R. [1973]. *Some efficient algorithms for solving systems of nonlinear equations*, SIAM Journal on Numerical Analysis, 10, 327-344.

Brown, K.M. [1966]. *A quadratically convergent method for solving simultaneous nonlinear equations*, PhD. Diss., Purdue U., Lafayette, Indiana.

Broyden, C. [1965]. *A class of methods for solving systems of nonlinear equations*, MAth. Comput. 19, 577-593.

Broyden, C., Dennis, J., Moré [1973]. *On the local and superlinear convergence of quasi-Newton Methods*, J.Inst.Math. & Applics., 12, 223-245.

Coleman, T., Moré, J. [1982]. *Software for estimating sparse Jacobian matrices*, ANL-82-37, Argonne National Laboratory, Argonne, Illinois.

Coleman, T. and Moré, J. [1983]. *Estimation of sparse Jacobian matrices and graph coloring problems*, SIAM Journal on Numerical analysis, 20, 187-209.

Curtis, A., Powell, M.J.D., Reid, J. [1974]. *On the estimation of sparse Jacobian matrices*, J. Inst. Math. Appl. 13, 117-119.

Dembo, R., Eisenstat, S., and Steihaug [1982]. *Inexact Newton Methods*, SIAM Journal on Numerical Analysis 19, 400-408.

Dennis, J., Gay, D., and Welsch, R., [1983]. *An adaptive nonlinear least-squares algorithm*, ACM Trans. Math. Softw. 7, 369-383.

Dennis, J. and Marwil, E. [1981]. *Direct secant updates of matrix factorizations*, TR 476, Dept. of Math. Sciences, Rice University, Houston, Texas.

Dennis, J. and More, J. [1974]
A characterization of superlinear convergence and its' application to quasi-Newton methods, Math. of Computation, 28, 549-560.

Gill, P. and Murray, W. [1978]. *Algorithms for the nonlinear least-squares problem*, SIAM Journal on Numerical Analysis, 15, 977-992.

Goldfarb, D. and Toint, P. [1983]. *Optimal estimation of Jacobian and Hessian matrices that arise in finite difference calculations*, Technical Report, Department of Industrial Engineering and Operations Research, Columbia University, New York.

Heath, M. [1982]. *Some extensions of an algorithm for sparse linear least squares problems*, SIAM Journal on Sci. Stat. Comput., 3.

Hebden, M.D. [1973]. *An algorithm for minimization using exact second derivatives*, Atomic Energy Research Establishment, Report T.P. 515, Harwell, England.

Levenberg, K. [1944]. *A method for the solution of certain nonlinear problems in least squares*, Quart. Appl. Math 2, 164-168.

Marwil, E.[1979]. *Convergence results for Schubert's method for solving sparse nonlinear equations*, SIAM Journal on Numerical Analysis 16, 588-604.

Marquardt, D.W. [1963]. *An algorithm for least squares estimation of nonlinear parameters*, SIAM J. Appl. Math. 11, 431-441.

Matula, D.W., Marble, G., and Isaacson, J.D. [1972]. *Graph coloring algorithms* in Graph Theory and Computing, R.C. Read, ed., Academic Press, New York, 104-122.

More, J. [1978]. *The Levenberg-Marquardt algorithm: Implementation and theory*, Proceedings of the Dundee Conference on Numerical Analysis, G.A. Watson, ed., Springer-Verlag.

More, J., Garbow, B., and Hillstrom, K.E. [1980]. *User guide for MINPACK-1*, Argonne National Laboratory, Report ANL-81-83, Argonne, IL.

Nazareth, L. [1980]. *Some recent approaches to solving large residual nonlinear least squares problems*, SIAM Review 22, 1-11.

Newsam, G., and Ramsdell, J. [1981]. *Estimation of sparse Jacobian matrices*, Aiken Computation Lab., TR-17-81, Harvard, Cambridge, Mass.

Powell, M.J.D. [1970]. *A hybrid method for nonlinear equations*, in Numerical Methods For Nonlinear Algebraic Equations, P. Rabinowitz, ed. Academic Press.

Rose, D.J. [1970]. *Triangulated graphs and the elimination process*, Journal of Mathematical Analysis and Applications, 32 597-609.

Schubert, L. [1970]. *Modification of a quasi-Newton method for nonlinear equations with a sparse Jacobian*, Math. Comp., 24, 27-30.

Speelpenning,B. [1980]. *Compiling Fast Partial Derivatives of Functions Given By Algorithms*, PhD Dissertation, Dept. of Computer Science, Urbana, Illinois.

Tarjan, R. [1975]. *Maximum cardinality search and chordal graphs*, unpublished notes.

Chapter 5   Large Unconstrained Optimization Problems

In this chapter we are interested in the problem

$$minimize \ f(x) \tag{5.1}$$

where $f:R^n \rightarrow R^1$ and f is twice continuously differentiable. We are particularily interested in the large sparse problem: the Hessian matrix, $\nabla^2 f(x)$, is large and sparse.

There are several recent books concerned with methods of optimization which provide suitable background material. I recommend Fletcher [1980] and Gill, Murray and Wright[1981], and Dennis and Schnabel[1983].

5.1 The Hessian Matrix

Much of the research effort, in unconstrained optimization, is concerned with the Hessian matrix. Most efficient algorithms require an approximation to $\nabla^2 f$; the 'acquisition' of the Hessian approximation and the solution of the corresponding linear system often represents the dominant work. There are three common ways of acquiring an approximation to $\nabla^2 f(x)$. Firstly, the second derivative functions may be coded and evaluated when needed. Secondly, $\nabla^2 f$ can be estimated by finite differences - we discuss this possibility in Sections 5.1.1, 5.1.2. Finally, a quasi-Newton approximation can be maintained - this approach is discussed in Section 5.2.

5.1.1 Direct Determination (Powell & Toint [1979], Coleman & More' [1982])

The problem of estimating a sparse Hessian matrix can be phrased as follows: Given the sparsity structure of a symmetric matrix A, obtain vectors $d_1, d_2, ..., d_p$ such that $Ad_1, Ad_2, \ldots, Ad_p$ determine A uniquely. In this section we are mainly concerned with direct methods for determining A based on partitions of the columns of A.

A *partition* of the columns of A is a division of the columns into groups $C_1, C_2, ..., C_p$ such that each column belongs to one and only one group. A partition of the columns of A is *consistent* with the direct determination of A if whenever a_{ij} is a nonzero element of A then the group containing column j has no other column with a nonzero in row i. A partition is *symmetrically consistent* if whenever a_{ij} is a nonzero element of A then the group containing column j has no other column with a nonzero in row i, or the group with column i has no other column with a nonzero in row j.

Given a consistent partition of the columns of A, it is straightforward to determine the elements of A with p evaluations of Ad by associating each group C with a direction d with components $\delta_j = 0$ if j does not belong to C, and

$\delta_j \neq 0$ otherwise. Then

$$Ad = \sum_{j \in C} \delta_j \, a_j$$

where $a_1, a_2, ..., a_n$ are the columns of A, and it follows that if column j is the only column in group C with a nonzero in row i then

$$(Ad)_i = \delta_j \, a_{ij} \, ,$$

and thus a_{ij} is determined. In this way, every nonzero of A is directly determined. In these notes we will not discuss the delicate numerical issues involving the appropriate value of δ_j: Gill, Murray, Saunders, and Wright [1983] have addressed this issue.

If A is symmetric, it is possible to determine A while only requiring that the partition be symmetrically consistent. Thus, given a symmetrically consistent partition of the columns of the symmetric matrix A, if column j is the only column in its group with a nonzero in row i then a_{ij} can be determined as above, while if column i is the only column in its group with a nonzero in row j then a_{ji} can be determined. Hence, every nonzero of A is directly determined with p evaluations of Ad. We are interested in partitions with the least number of nonzeroes because each group represents one gradient evaluation.

Partition Problem: Obtain a symmetrically consistent partition of the columns of the symmetric matrix A with the fewest number of groups.

In the symmetric case, the appropriate graph is the graph $G_S(A)$ with vertex set $\{ a_1, a_2, ..., a_n \}$ and edge (a_i, a_j) if and only if $i \neq j$ and $a_{ij} \neq 0$. In graph theory terminology, $G_S(A)$ is the *adjacency graph* of the symmetric matrix A.

A coloring ϕ of $G_S(A)$ does not necessarily induce a symmetrically consistent partition of the columns of a symmetric matrix A; it is necessary to restrict the class of colorings.

Definition. A mapping $\phi : V \rightarrow \{ 1, 2, ..., p \}$ is a symmetric p-coloring of a graph $G = (V, E)$ if ϕ is a p-coloring of G and if ϕ is not a 2-coloring for any path in G of length 3. The symmetric chromatic number $\chi_\sigma(G)$ is the smallest p for which G has a symmetric p-coloring.

A *path in G* of length l is a sequence $(v_0, v_1, ..., v_l)$, of distinct vertices in G such that v_{i-1} is adjacent to v_i for $1 \leq i \leq l$. Thus, if ϕ is a symmetric p-coloring of G then the situation

      Red   Blue   Red   Blue

is not allowed.

We can now express the partition problem for symmetric matrices as a graph coloring problem.

Theorem 5.1.1 Let A be a symmetric matrix with nonzero diagonal elements. The mapping ϕ is a symmetric coloring of $G_S(A)$ if and only if ϕ induces a symmetrically consistent partition of the columns of A.

In view of Theorem 5.1.1 the partition problem is equivalent to the following problem.

Symmetric Graph Coloring Problem: Obtain a minimum symmetric coloring of $G_S(A)$.

Coleman and Moré[1982] established that the symmetric graph coloring problem is just as hard as the general graph coloring problem. Therefore, a fast heuristic algorithm is in order. Coleman and Moré discussed several possibilities.

5.1.2 Substitution Methods

Coleman and Moré proved that a direct method based on an optimal symmetric partition has no advantage over an direct method based on an optimal (unsymmetric) partition when the matrix is a *symmetric band matrix*. This suggests that direct symmetric methods may not be able to take full advantage of the symmetry of a general matrix. In this section we explore a type of indirect method which is able, in particular, to produce the desired results for banded matrices.

The direct method described in Section 5.1.1 can be viewed as inducing a diagonal linear system where the unknowns are the nonzeroes of A. In order to exploit the symmetry of A to a greater degree, we can broaden the class of induced systems. Of course it is essential that the systems be easy to solve and involve no (or at least simple) factorizations. These constraints lead us to consider triangular systems. We will call methods that induce triangular linear systems, *substitution methods*. If we further restrict ourselves to methods based on a *partition* of columns, then we are faced with the following question: What class of graph colorings characterize substitution methods? (Recall: symmetric colorings characterize direct methods.)

Before we can answer this question it is necessary to formally define a substitution method based on a partition of columns. Let $C = C_1, \ldots, C_t$ be a partition of the columns of the symmetric matrix A. Let N denote the set of (i,j) pairs of indices of the nonzeroes of A. C defines a substitution method if there exists an ordering $\pi = <(i(k),j(k))>$ of N and a corresponding sequence of sets $S_k = S_{k-1} \bigcup \{(i(k),j(k))\}$ with the following properties.

1. $S_0 = \{\}$
2. *EITHER* column $j(k)$ belongs to a group, C_r say, such that if $l \epsilon C_r$, $l \neq j(k)$, and $(i(k),l) \epsilon N$ then $(i(k),l) \epsilon S_{k-1}$,

   OR column $i(k)$ belongs to a group, C_s say, such that if $l \epsilon C_s$, $l \neq i(k)$, and $(j(k),l) \epsilon N$ then $(j(k),l) \epsilon S_{k-1}$
3. $S_{|N|} = N$.

It is clear that A can be determined via back substitution since $A(i(k),j(k))$ depends on $A(i(h),j(h))$, $h < k$. Here is a clean characterization of a substitution method based on a partition of columns.

Theorem 5.1.2: Let ϕ be a coloring of $G_S(A)$. ϕ corresponds to a substitution method if and only if ϕ does not contain a bi-colored cycle.

Proof: Suppose that ϕ contains a bi-colored cycle using colors RED and BLUE. It is clear that the off-diagonal elements in the RED columns depend on the off-diagonal columns in the BLUE columns and vice-versa. Hence a sequence satisfying property 2 is impossible.

Suppose that ϕ does not contain a bi-colored cycle. Then, for every pair of colors ϕ_i, ϕ_j, consider the induced forest F_{ij}. It is clear that for every maximum tree in F_{ij} a suitable ordering exists by recursively ordering the 'free' edges first (the ordering is not unique) and the proof is complete

The remaining task is to efficiently produce low cardinality graph colorings with this cycle property and then retrieve the Hessian information. The full generality of this characterization has not been investigated to date: the heuristic procedures described later in this section take a narrower point of view. Before discussing these procedures, a comment about growth of error is in order.

The substitution methods defined above implicitly solve a very large triangular system of equations (of order τ, where τ is the number of nonzeroes in the Hessian matrix). This increase in arithmetic (over diagonal systems) will result in increased growth of error. Powell and Toint [1979] claimed that for a *particular* class of substitution methods (discussed later in this section), dangerous growth of error is not generally a concern since they proved that each unknown in the Hessian matrix is dependent on *at most* n other unknowns. Fortunately, this same claim can be made for *any* substitution method based on a partition of columns.

Theorem 5.1.3: For any substitution method based on a partition of columns, each unknown element of A is dependent on at most n other elements.

Proof: The elements of A can be determined by considering each forest F_{ij}, in turn. But each forest has at most n edges and the result follows.

Lower triangular substitution methods form a subclass of the class of substitution methods discussed above. They were first introduced by Powell and Toint [1979]. Let A be a symmetric matrix and let L be the lower triangular part of A; that is, L is a lower triangular matrix such that $A - L$ is strictly upper triangular. A lower triangular substitution method is based on the result of Powell and Toint [1979] that if $C_1, C_2, \ldots, C_p$ is a consistent partition of the columns of L then A can be determined indirectly with p evaluations of Ad. It is not difficult to establish this result. With each group C associate a direction d with components $\delta_j \neq 0$ if j belongs to C and $\delta_j = 0$ otherwise. Then

$$Ad = \sum_{j \in C} \delta_j a_j$$

where $a_1, a_2, \ldots, a_n$ are the columns of A. To determine a_{ij} with $i \geq j$ note that if column j is the only column in group C with a nonzero in row $i \geq j$ then

$$(Ad)_i = \delta_j a_{ij} + \sum_{l>i, l \in C} \delta_l a_{li} . \tag{5.1.1}$$

This expression shows that a_{ij} depends on $(Ad)_i$ and on elements of L in rows $l > i$. Thus L can be determined indirectly by first determining the n-th row of L and then solving for the remaining rows of L in the order $n-1, n-2, \ldots, 1$. Another consequence of (5.1.1) is that computing a_{ij} requires, at most, ρ_i operations where ρ_i is the number of nonzeroes in the i-th row of A. Thus computing all of A requires less than

$$\sum_{i=1}^{n} \rho_i^2$$

arithmetic operations, and this makes a triangular substitution method attractive in terms of the overhead. On the other hand, computing all of A with a direct method requires about τ arithmetic operations where τ is the number of nonzeroes in A. Another difference between direct methods and triangular substitution methods is that in a triangular substitution method (indeed in any substitution method) the computation of a_{ij} requires a sequence of substitutions which may magnify errors considerably, while in a direct method there is no magnification of errors. Powell and Toint [1979] showed that magnification of errors can only occur when the ratio of the largest to the smallest component of d is large.

Powell and Toint [1979] also noted that the number of groups in a consistent partition of the columns of L depends on the ordering of the rows and columns of A. Thus, if π is a permutation matrix and L_π is the lower triangular part of $\pi^T A \pi$ then we may have

$$\chi(G_U(L_\pi)) < \chi(G_U(L)).$$

For example, if A has an arrowhead structure, then it is possible to choose the permutation π so that the chromatic number of $G_U(L_\pi)$ is any integer in the interval $[2, n]$. Since Powell and Toint were unaware of the existence of the smallest-last ordering in the graph theory literature, it is interesting to note that the algorithm proposed by Powell and Toint [1979] for choosing the permutation matrix

π is the smallest-last ordering of $G_S(A)$. There are good reasons for choosing this ordering to define the permutation matrix π; see Powell and Toint and Coleman and Moré for further details.

In graph theory terminology, the lower triangular substitution method of Powell and Toint consists of choosing the smallest-last ordering for the vertices of $G_S(A)$ and then coloring $G_U(L_\pi)$ with a sequential coloring algorithm. If column j is in position $\pi(j)$ of the smallest-last ordering of $G_S(A)$ then the permutation matrix π can be identified with the smallest-last ordering by setting the j-th column of π to the $\pi(j)$-th column of the identity matrix. Thus the vertices of $G_U(L_\pi)$ have the induced ordering

$$a_{\pi(1)}, a_{\pi(2)}, \ldots, a_{\pi(n)}. \tag{5.1.3}$$

Powell and Toint used the sequential coloring algorithm with this ordering to color $G_U(L_\pi)$. There is no compelling reason for using the sequential coloring algorithm with this ordering and, in fact, numerical results show that the use of other orderings in the sequential coloring algorithm tends to reduce the number of evaluations of Ad needed by the triangular substitution method.

To further explore the properties of triangular substitution methods, we characterize a coloring of $G_U(L_\pi)$ as a restricted coloring of $G_S(A)$ in the following sense.

Definition. A mapping $\phi: V \rightarrow \{1, 2, \ldots, p\}$ is a triangular p-coloring of a graph $G = (V, E)$ if ϕ is a p-coloring of G and if there is an ordering $v_1, v_2, \ldots, v_n$ of the vertices of G such that ϕ is not a 2-coloring for any path (v_i, v_k, v_j) with $k > \max(i, j)$. The triangular chromatic number $\chi_r(G)$ of G is the smallest p for which G has a triangular p-coloring.

For some graphs it is not difficult to determine $\chi_r(G)$. For example, if G is a band graph with bandwidth β then $\chi_r(G) = 1 + \beta$. Band graphs thus show that the triangular chromatic number of a graph can be considerably smaller than the symmetric chromatic number of a graph. Our next result shows that the triangular chromatic number of $G_S(A)$ is the smallest number of evaluations of Ad needed to determine a symmetric matrix A with a triangular substitution method.

Theorem 5.1.4 *Let A be a symmetric matrix with nonzero diagonal elements. The mapping ϕ is a triangular coloring of $G_S(A)$ if and only if ϕ is a coloring of $G_U(L_\pi)$ for some permutation matrix π.*

Proof: It is sufficient to show that ϕ is a triangular coloring of $G_S(A)$ with the ordering $a_1, a_2, \ldots, a_n$ of the vertices of $G_S(A)$ if and only if ϕ is a coloring of $G_U(L)$.

First assume that ϕ is a triangular coloring of $G_S(A)$ and let (a_i, a_j) be an edge of $G_U(L)$. Then (a_i, a_j) is an edge of $G_S(A)$ or there is an index $k > \max(i, j)$ such that (a_i, a_k) and (a_k, a_j) are edges of $G_S(A)$. Since ϕ is a triangular coloring of $G_S(A)$, we must have that $\phi(a_i) \neq \phi(a_j)$. Hence ϕ is a coloring of $G_U(L)$.

Now assume that ϕ is a coloring of $G_U(L)$. Then ϕ is a coloring of $G_S(A)$ because $\bar{G_S}(A)$ is a subgraph of $G_U(L)$ whenever A has nonzero diagonal elements. If (a_i, a_k, a_j) is a path in $G_S(A)$ with $k > \max(i,j)$ then (a_i, a_j) is an edge of $G_U(L)$ and hence $\phi(a_i) \neq \phi(a_j)$. Thus ϕ is not a 2-coloring of (a_i, a_k, a_j).

An important consequence of Theorem 5.1.4 is that it shows that triangular substitution methods are implicitly trying to solve a restricted graph coloring problem.

Triangular Graph Coloring Problem: Obtain a minimum triangular coloring of $G_S(A)$.

Coleman and Moré [1982] established that this restricted coloring problem is just as difficult as the general graph coloring problem - thus the use of efficient heuristics is justified.

5.2 Sparse Quasi-Newton Methods

Quasi-Newton methods, which involve low-rank corrections to a current approximate Hessian matrix, have played a major role in the development of efficient algorithms for dense unconstrained minimization. There has been considerable recent effort to adapt these methods to the large sparse setting by preserving sparsity. This effort has met with only partial success. In these notes we will very briefly review the dense case (Dennis and Moré [1977] provide an excellent discussion of the theory behind quasi-Newton methods); we will then discuss attempts to extend the methods to the large sparse setting.

The Dense Updates

Let x and B represent the current point and Hessian approximation respectively. If the new point is $x^+ = x + s$ then the new Hessian approximation is required to satisfy the quasi-Newton equation,

$$B^+ s = y, \quad \text{where } y := \nabla f(x^+) - \nabla f(x), \tag{5.2.1}$$

and a symmetry condition,

$$B^+ = (B^+)^T \tag{5.2.2}$$

In addition, it is usually required that $\|B^+ - B\|$ be restricted; finally, a useful property is that B^+ inherit positive-definiteness.

For example, if we require that B^+ solve the optimization problem,

$$\min\{\|\bar{B} - B\|_F : \bar{B} \text{ symmetric}, \bar{B}s = y\} \tag{5.2.3}$$

then B^+ is defined by the Powell-Symmetric-Broyden update (PSB):

$$B^+ = B + \frac{(y - Bs)s^T + s(y - Bs)^T}{s^T s} - \frac{[(y - Bs)^T s]ss^T}{(s^T s)^2} \tag{5.2.4}$$

Another possibility is to use a weighted Frobenius norm. For example, if W is a positive definite matrix let B^+ be defined by the optimization problem

$$\min \{\| W^{-\frac{1}{2}} (\bar{B}-B) W^{-\frac{1}{2}} \|_F, \ \bar{B} \ symmetric, \ \bar{B}s = y\} \tag{5.2.5}$$

The solution to (5.2.5) is given by

$$B^+ = B + \frac{1}{2} (W\lambda s^T W + Ws\lambda^T W) \tag{5.2.6}$$

where λ is a vector of Lagrange multipliers. A reasonable choice for W is $\nabla^2 f(x^*)$ - unfortunately, $\nabla^2 f(x^*)$ is unknown. A compromise is to let W be a positive definite matrix satisfying $Ws = y$. In this case the solution to (5.2.5) simplifies to the DFP update

$$B^+ = B + \frac{(y-Bs)y^T + y(y-Bs)^T}{y^T s} - \frac{[s^T(y-Bs)]yy^T}{(y^T s)^2} \tag{5.2.7}$$

An important consequence of this update is that if B is positive definite and $y^T s > 0$ then B^+ is positive definite.

In a similar vein, the optimization problem

$$\min\{\| W^{\frac{1}{2}} (\bar{B}^{-1} - B^{-1}) W^{\frac{1}{2}} \|_F, \ \bar{B} \ symmetric, \ \bar{B}s = y\} \tag{5.2.8}$$

where W is a positive definite matrix satisfying $Ws = y$, is solved by the BFGS update

$$(B^+)^{-1} = B^{-1} + \frac{(s-B^{-1}y)s^T + s(s-B^{-1}y)}{s^T y} - \frac{[y^T(s-B^{-1}y)]ss^T}{(s^T s)^2} \tag{5.2.9}$$

Equivalently,

$$B^+ = B + \frac{yy^T}{(s^T y)} - \frac{Bss^T B}{s^T Bs} \tag{5.2.10}$$

If B is positive definite and $s^T y > 0$ then B^+ is positive definite.

The three updates mentioned above have similar asymptotic convergence properties: under reasonable assumptions they exhibit superlinear convergence. In practise, they differ markedly - the BFGS and DFP updates are clearly superior to the PSB update (also, the positive definite inheretance allows for a more efficient implementation). The BFGS update typically outperforms the DFP rule.

The three updates described above are all unsuitable for large sparse problems since they will cause complete fill, in general.

Sparse PSB (Toint[1977], Marwil[1978])

A reasonable and practical goal is to have the sparsity of B reflect the sparsity of $\nabla^2 f$. Therefore it is natural to extend the Powell-Symmetric-Broyden update by requiring that $B^+ \in S(\nabla^2 f)$, where $S(\nabla^2 f)$ is the space of matrices with the sparsity of $\nabla^2 f$. The corresponding optimization problem, which defines B^+, is

$$\min \{\|\bar{B}-B\|_F, \ \bar{B} \in Q(y,s) \cap S(\nabla^2 f)\} \tag{5.2.11}$$

where $Q(y,s)$ is the affine set of *symmetric* matrices which send $s \to y$. If we define $C := \bar{B} - B = \frac{1}{2}(E + E^T)$ where E is a matrix (unknown) which need not be symmetric, and $r := y - Bs$, then problem (5.2.11) becomes

$$\min \frac{1}{8}\|E + E^T\|_F^2, \ (E_{i\bullet} + E_{\bullet i})s_{[i]} = 2r_i, \ \text{for } i=1,...,n$$

where $E_{i\bullet}(E_{\bullet i})$ is row (column) i of E and $s_{[i]}$ is the orthogonal projection of s onto S_i (the space of n-vectors with the sparsity pattern of row i of $\nabla^2 f$).

The Lagrangian function of this optimization problem can be written

$$\phi(E,\lambda) = \frac{1}{8}\sum_{i=1}^{n}\sum_{j=1}^{n}(E_{ij}^2 + E_{ji}^2 + 2E_{ij}E_{ji}) - \sum_{i=1}^{n}\lambda_i\{(E_{i\bullet} + E_{\bullet i})s_{[i]} - 2r_i\}.$$

Differentiating ϕ with respect to E_{ij} and setting the result to zero yields

$$\frac{1}{2}(E_{ij} + E_{ji}) - \lambda_i s_{[i]_j} - \lambda_j s_{[j]_i} = 0$$

or, equivalently

$$C = \sum_{i=1}^{n}\lambda_i[e_i(s_{[i]})^T + s_{[i]}e_i^T]$$

The above equation can be written as $Q\lambda = r$ where

$$Q = \sum_{i=1}^{n}[s_{[i]}s_i + \|s_{[i]}\|_2^2 e_i]e_i^T$$

The matrix Q is symmetric and has the same sparsity pattern as B. In addition, Q is positive definite if $\|s_{[i]}\| > 0$ for all i. If some of the vectors $s_{[i]}$ are zero, then the size of the problem is reduced by setting the corresponding row and column of C to zero. In summary, the first step is to form Q and solve $Q\lambda = r$, for λ. The Hessian matrix is then updated $B \leftarrow B + C$, where C is the symmetric rank-n correction defined above.

Toint[1977] has established that the sparse PSB is locally and superlinearly convergent, under reasonable assumptions. However, the performance of this method in practise has not been outstanding - it seems best used as an updating technique between frequent evaluations of the true Hessian matrix.

Since the BFGS and DFP updates outperform PSB in the dense situation, it seems reasonable to try and adapt these methods to the sparse setting.

Sparse DFP (BFGS)

A natural way to try and adapt the DFP update is by considering the optimization problem (5.2.8) with the extra constraint $\bar{B} \in S(\nabla^2 f)$. (A similar remark can be made about the BFGS update.) Unfortunately, a closed form computable solution to this problem is not known - the difficulty rests with the fact that for a general symmetric matrix W, WAW does not have the sparsity pattern of A

(even if W does).

The solution to problem (5.28) with the extra constraint $B \in S(\nabla^2 f)$ can be written as

$$B^+ = P_W^{A \cap s}(B)$$

wher $P_W^Y(M)$ projects the matrix M, W-orthogonally, onto the manifold $Y(M)$. Dennis and Schnabel[1979] proved that the solution can be written

$$B^+ = B + \frac{1}{2} P_W^S[W \lambda s^T W + W s \lambda^T W]$$

where λ is defined by the system $B^+(\lambda)s = y$. Unfortunately, we do not know how to efficiently compute P_W^S or $P_W^{A \cap s}$ for a general matrix W.

The Method of Shanno[1980] and Toint[1981a]

A computable compromise can be achieved in the following way. Firstly, let $\hat{B}$ solve the dense problem:

$$\min\{\| W^{-\frac{1}{2}}(\bar{B} - B) W^{-\frac{1}{2}} \|, \ \bar{B} \ symmetric, \ \bar{B}s = y\}$$

Secondly, let $\tilde{B}$ be the nearest matrix to $\hat{B}$ (in the Frobenius norm). That is $\tilde{B}$ solves

$$\min\{\|\bar{B} - \hat{B}\|_F, \ \bar{B} \in S(\nabla^2 f)\}$$

Note that the solution is trivially computed by 'zeroing' the appropriate positions of $\hat{B}$. Finally, the matrix B^+ is determined: B^+ is the nearest matrix to $\tilde{B}$, in the Frobenius norm, which also satisfies $B^+ s = y$. That is, B^+ solves

$$\min\{\|\bar{B} - \tilde{B}\|_F, \bar{B} \ symmetric, \ \bar{B}s = y, \ \bar{B} \in S(\nabla^2 f)\} \qquad (5.2.12)$$

But this optimization problem has exactly the same form as (5.2.11) and therefore a solution can be achieved by the procedure described in the previous section (due to Toint[1977]). Computationally, it is important to note that it is never necessary to compute the dense matrix $\hat{B}$. Thapa[1979] was the first to observe this.

In summary, the compromise solution assigns

$$B^+ \leftarrow P_I^{A \cap s}[P_W^A(B)] = P_I^{A \cap s}\{P_I^S[P_W^A(B)]\}$$

where $P_W^Y(M)$ projects M (W-orthogonally) onto Y, A is the affine set of symmetric matrices satisfying $Ms = y$, and S is the linear space of matrices with the sparsity of $\nabla^2 f$. (Note that the 'ideal' solution is $B^+ \leftarrow P_W^{A \cap s}(B)$).

Shanno has reported good numerical results with the use of this update. However, Powell and Toint[1981] indicated that the update can permit exponential growth in error of the second derivative approximations. The method may not exhibit superlinear convergence. In addition, the matrix B^+ will not necessarily be positive definite even if B is positive definite and $y^T s > 0$.

It is interesting to note that this technique is applicable to any dense update: Suppose that $\overline{C}$ represents an arbitrary *dense* update: $\hat{B} \leftarrow B + \overline{C}$. Then a sparse matrix B^+, which satisfies the quasi-Newton equation, can be obtained in the following way:

$$B^+ = P_I^{A \cap s}(\hat{B}) = P_I^{A \cap s}(P_I^S \hat{B})$$

Unfortunately, the ideas discussed above do not lead to positive definite updates. Is it possible to achieve positive definiteness in addition to other 'desirable' properties ?

Desirable Properties

Ideally, the symmetric matrix B^+ would satisfy

1. B^+ is positive definite,
2. $B^+ s = y$,
3. $B^+ \epsilon S(\nabla^2 f)$.

It is impossible to always attain all three desirable properties. For example, suppose that sparsity constraints force B^+ to be diagonal, say $B^+ = diag(d_1, d_2)$. If $s = (2,-1)$ and $y = (1,1)$ then condition 2 implies $d_2 < 0$ while condition 1 implies $d_2 > 0$. At first this example seems quite discouraging, however Toint[1981b] has proven that the *only* situation in which an update satisfying properties 1–3 does not exist is when B can be decomposed into block diagonal form, provided $s_i \neq 0$, $i=1,...,n$. Let $G_S(B)$ denote the adjacency graph of B.

Theorem 5.2.1 *If $G_S(B)$ is a connected graph, $y^T s > 0$ and $s_i \neq 0$ for $i=1,...,n$ then there is a positive definite matrix B^+ satisfying properties 1 - 3.*

Proof: A full proof is given by Toint[1981b]. Here we provide a sketch. Let C be any matrix such that $B + C$ satisfies sparsity, symmetry and the quasi-Newton equation. The basic idea is to show that a matrix E exists such that
  i) $E = E^T$,
  ii) $E \epsilon S(B)$
  iii) $Es = 0$,
  iv) if $z \neq 0$ and $z \neq \lambda s$, for some λ, then $z^T E z > 0$.
Note that $s^T (B + C + \sigma E) s = s^T y > 0$, for any σ. It is not difficult to show that the matrix $B + C + \sigma E$ satisfies the *three* desirable properties for σ sufficiently large.

We now show that a matrix E, satisfying properties i) - iv), can be constructed. Define $E_{ij} = -s_{[i]} s_{[j]}$, if $i \neq j$ and $E_{ii} = \|s_{[i]}\|_2^2 - s_i^2$. Properties i),ii), and iii) are easy to establish. In order to prove iv), note that E can be written

$$E = \sum_{i < j, \, (i,j) \, \epsilon \, T} R_{ij}$$

where T is the set of *nonzero* positions of B, and R_{ij} is the zero matrix with the exception that the (i,j) submatrix is $\begin{bmatrix} s_j^2 & -s_i s_j \\ -s_i s_j & s_i^2 \end{bmatrix}$. The matrix R_{ij} is a rank-one matrix with the following structure: $n-2$ of the zero eigenvalues have eigenvectors z with

$$z_l = 0, \quad l = i, j \tag{5.2.13}$$

Another zero eigenvalue has eigenvector v_1^{ij}:

$$(v_1^{ij})_l = \begin{cases} s_i & l=i \\ s_j & l=j \\ 0 & otherwise \end{cases}$$

Finally, R_{ij} has a positive eigenvalue $(s_i^2 + s_j^2)$ with eigenvector v_2^{ij} defined by

$$(v_2^{ij})_l = \begin{cases} 1 & l=i \\ \dfrac{-s_i}{s_j} & l=j \\ 0 & otherwise \end{cases}$$

Therefore R_{ij} is positive semi-definite and hence E is also.

If $z \neq 0$ then for all $(i,j) \in E$, $i < j$,

$$z = z^{ij} + \alpha_1^{ij} v_1^{ij} + \alpha_2^{ij} v_2^{ij} \tag{5.2.14}$$

where z^{ij} is the orthogonal projection of z onto the space of vectors satisfying (5.2.13).

We now show that for any vector z which is not a multiple of s there is at least one R_{ij} such that $z^T R_{ij} z > 0$: Property $iv)$ will then be established. But note that if u and t are indices such that $(0,...,s_u,0,...,s_t,0,...,0)$ and $(0,...,z_u,0,...,z_t,0,...,0)$ are linearly independent then $\alpha_2^{ut} \neq 0$ and

$$z^T R_{ut} z = (\alpha_2^{ut})^2 (s_u^2 + s_t^2) \|v_2^{ut}\|^2$$

$$= (\alpha_2^{ut})^2 (\frac{s_u^2 + s_t^2}{s_t})^2 > 0 \tag{5.2.15}$$

It only remains to show that for any vector z which is not a multiple of s there exists an index pair $(u,t) \in T$ with $u < t$ such that (s_u, s_t) and (z_u, z_t) are linearly independent. Since $G_S(B)$ has exactly one component, it follows that there is a connected path $\pi: (x_{i_1}, \ldots, x_{i_p})$ which includes every vertex at least once: that is, $\{1,2,...,n\} \subset \{i_1, \ldots, i_p\}$. If z_{i_1} is zero, let i_q be the *first* integer in π such that z_{i_q} is nonzero. Otherwise, let i_q be the *first* integer in π such that $s_{i_q} \neq z_{i_q}$. (Such a q exists since otherwise $z = s$.) Let

$$(u,t) = \begin{cases} i_{q-1}, i_q & \text{if } i_{q-1} < i_q \\ i_q, i_{q-1} & \text{if } i_{q-1} > i_q \end{cases}$$

Clearly $(u,t) \in T$ and $u < t$. Suppose $(s_u, s_t) = \lambda(z_u, z_t)$ for some λ. By assumption, no component of s is zero; hence, $z_u \neq 0$ and $z_t \neq 0$ and $\lambda \neq 0$. But either $\{z_u = s_u$ and $s_t \neq z_t\}$ or $\{z_u \neq s_u$ and $s_t = z_t\}$. Either case immediately contradicts the linear dependence assumption .

Theorem 5.2.1 is encouraging: it suggests that in most cases a matrix satisfying the desirable properties exists. Toint[1982b] also argued that the assumption $s_i \neq 0$ can be dropped provided a 'reduced' problem is solved. In addition, a problem with a disconnected adjacency graph can be handled in a separable fashion.

Unfortunately, Sorensen[1981] has demonstrated that there are potential problems even if the assumptions of Theorem 5.2.1 are satisfied. In particular, a *small* value of s_i in combination with sparsity constraints can result in an arbitrarily large growth of error in the Hessian approximation. This strongly suggests that a stable sparse update will probably necessitate the relaxation of at least one of the desirable properties.

5.3 Sparse Quasi-Newton Methods For Structured Problems

As was demonstrated in the previous section, general sparse updating strategies with attractive theoretical and numerical properties are hard to come by.
Therefore, it may be worthwhile designing specific sparse updating strategies for
specific common structures. The following ideas were proposed by Griewank and
Toint[1982].

Suppose that $f(x)$ can be written as

$$f(x) = \sum_{i=1}^{m} f_i(x) \tag{5.3.1}$$

where each element function f_i depends on only $\eta_i \ll n$ components of f. For
example, the following overlapping structure may represent $\nabla^2 f$ where square i
corresponds to $\nabla^2 f_i$.

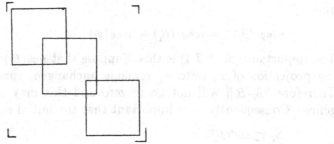

Let us assume that (5.3.1) defines a convex decomposition: each element function
f_i is convex ($\nabla^2 f_i(x)$ is positive semi-definite). Further, assume that $\nabla^2 f(x^*)$ is
positive definite. Finally, we assume that the nullspace of each element Hessian
$\nabla^2 f_i$ *does not depend on* x. (These assumptions are satisfied for certain finite element applications.)

Let H_i and B_i represent the reduced (η_i *by* η_i) Hessian matrix and approximation, with respect to f_i, respectively. Similarly, let x_i, s_i and y_i represent the
reduced versions, with respect to f_i, of x, s, and y. The nullspace of H_i, with
respect to x_i, is denoted by N_i while the orthogonal complement is denoted by
N_i'. ($|N_i'| + |N_i| = \eta_i$).

It is important to note that s_i is not directly related to B_i since s_i is
extracted from the solution of $Bs = -\nabla f$, *where* B *is the 'sum' of the matrices* H_i.
Hence it is not feasible to ensure $s_i^T y_i > 0$ via a line search.

It is reasonable to require that

$$B_i^+ s_i = y_i, \quad i=1,...,m. \tag{5.3.2}$$

In addition, since H_i is positive semi-definite, it is appropriate to maintain a positive semi-definite approximation B_i. Finally, since N_i is invariant with x, it is
natural to try and obtain

$$null(B_i^+) = N_i \tag{5.3.3}$$

Note that

$$y_i = \bar{H}_i\, s_i, \text{ where } \bar{H}_i = \int_0^1 H_i(x_i + s_i\, t)\, dt$$

and hence $y_i \in N_i'$. Therefore, condition (5.3.3) will not compromise (5.3.2). Finally, the average curvature over s_i will always be non-negative since $y_i = \bar{H}_i\, s_i \Rightarrow y_i^T s_i \geq 0$.

The DFP Update

The DFP update is not defined if $y_i^T s_i = 0$; however, it is not hard to see that if $y_i^T s_i > 0$ and B_i is positive semi-definite then B_i^+ will be positive semi-definite. Furthermore, one can show that

$$y_i^T z = 0 \Rightarrow z^T B_i^+ z = z^T B_i z \tag{5.3.4}$$

and

$$range(B_i^+) = range(B_i) + span(y_i) \tag{5.3.5}$$

The importance of (5.3.4) is this: Suppose that $null(B_i)$ does *not* contain N_i. Then the projection of B_i onto N_i remains unchanged, considering (5.3.4) and $y_i \in N_i'$. Therefore $\|B_i - \bar{B}_i\|$ will not go to zero and this may prevent superlinear convergence. Consequently, it is important that the initial approximation, B_i^0, satisfy

$$N_i \subset null(B_i^0) \tag{5.3.6}$$

For example, B_i^0 can be initialized to the zero matrix. Equation (5.3.5) implies that the rank of B_i will increase at every iteration unless $y_i \in range(B_i)$. Therefore, it is reasonable to expect that after a number of iterations, $null(B_i) = N_i$, provided condition (5.3.6) is satisfied. A major computational difficulty with this approach is that the resulting summation matrix B may be singular in the initial iterations. Therefore it will be difficult to determine the quasi-Newton direction.

The BFGS Update

The BFGS update is well-defined if $y_i^T s_i > 0$ and $s_i^T B_i\, s_i > 0$ (equivalently, $B_i\, s_i \neq 0$). If these conditions are satisfied then it is not hard to show that B_i^+ will inherit positive semi-definiteness; in addition,

$$rank(B_i^+) = rank(B_i) \tag{5.3.7}$$

Condition (5.3.7) suggests that if $rank(B_i^0) \neq n_i - dim(N_i)$ then the BFGS update will not self-correct and identify the correct nullspace. This is true, however if $rank(B_i) \geq n_i - dim(N_i)$ then things will work out, asymptotically, since

$$z^T B_i^+ z = z^T B_i z - \frac{(z^T B_i\, s_i)^2}{s_i^T B_i\, s_i} \tag{5.3.8}$$

if $y_i^T z = 0$. This relationship, in conjunction with $y_i \in N_i'$, indicates that the projection of B_i onto N_i will approach zero in the limit. (Of course, if $B_i z = 0$ for z in N_i then $B_i^+ z = 0$.)

The BFGS procedure does not suffer from potential singularity problems, as is the case with the DFP update, since low rank approximations to H_i in the initial stages are not required. This is because (5.3.8) suggests that it is preferable to start with

$$null(B_i) \subset N_i \qquad (5.3.9)$$

if it cannot be ensured that $null(B_i) = N_i$: Hence it is better to over-estimate the rank. (Notice that if $null(H_i)$ varies with x, or N_i is not known beforehand, then (5.3.8) suggests that the BFGS update can still be expected to yield good approximations - the same cannot be said for the DFP update - see Griewank and Toint[1982] for more details).

An implementation of these ideas must pay careful attention to the maintenance of numerical positive semi-definiteness. Griewank and Toint[1982] noted that if a p by n_i matrix Q can be supplied where $N_i = range(Q^T)$ and $p = dim(N_i)$ then the implementation can be simplified. In particular, the variables of f_i can be permuted so that the last p columns of Q are nonsingular: Therefore, without loss of generality we can assume that $Q = (Q_0, I)$ where Q_0 is p by $n_i - p$. Note that a basis for the range space is given by $(I, Q_0^T)^T$. It is not hard to show that

$$B_i = \begin{bmatrix} I \\ Q_0 \end{bmatrix} (B_i)_{11} (I, Q_0^T) \qquad (5.3.10)$$

where $(B_i)_{11}$ is the leading (positive-definite) submatrix of B_i of order p_i. It is then possible to update the Cholesky factor of $(B_i)_{11}$ and maintain numerical positive definiteness of $(B_i)_{11}$. This approach is especially useful if the quasi-Newton direction is solved for using an iterative technique.

Outstanding Problems

Numerical results by Griewank and Toint[1982] indicate that these ideas can be successfully applied. Further work along these lines is needed. In particular, stable implementation techniques for maintaining semi-positive definiteness need further study. In addition, it is not clear that the nonconvex case can be successfully handled: Griewank and Toint[1983] proposed a 'juggler's solution' which needs further experimentation. Finally, is it possible to decompose a general sparse problem, automatically and efficiently, into a sum of 'small' element functions?

5.4 Conjugate Gradient Methods

Congugate gradient approaches to the large unconstrained optimization problem are attractive because, in their raw form, they require only the storage of several n-vectors. Of course in order to improve their performance it may be necessary to compute and store a *pre-conditioning matrix* also. It is usual to divide conjugate gradient methods into 2 classes: linear and nonlinear. We follow that convention here.

5.4.1 The Linear Conjugate Gradient Approach

The linear conjugate gradient algorithm can be applied to the Newton equations when the Hessian matrix is positive definite. It is not necessary to store the matrix H_k since a product $H_k p$ can be estimated by a single finite difference. (See Section 1.3 for an algorithmic description of the linear conjugate gradient method.) On the other hand, it may be more efficient to estimate the entire sparse matrix H_k by finite differences if H_k can be stored. In particular, if the number of iterations required to solve the linear system to suitable accuracy is greater than t, when H_k can be estimated in t differences, then it may be preferable to estimate H_k by differencing and then compute the matrix-vector products $H_k p$.

As discussed in Section 4.2, it is not necessary to solve the linear systems exactly to achieve a superlinear convergence rate. Application of this notion can result in considerable savings. In addition there are several techniques for dealing with an indefinite matrix within a conjugate gradient framework. O'Leary[1980] and Paige and Saunders[1982] proposed Lanczos-related solutions. Steihaug[1983] suggested computing conjugate directions, in the usual fashion, until a direction of negative curvature (indicating indefiniteness) is found: the CG procedure is then aborted and the direction of negative curvature is used in conjunction with the previous combination of conjugate directions.

In the next section we briefly describe an implementation, due to Steihaug[1983], of the linear conjugate gradient algorithm for large unconstrained optimization problems.

A Global CG Algorithm (Steihaug[1983])

The dogleg algorithm for nonlinear least squares is described in Section 4.4.4. This idea can readily be adapted to unconstrained optimization (Powell [1970]). In particular, let

$$\Psi(s) := \nabla f(x)^T s + \tfrac{1}{2}\, s^T B s \tag{5.4.1}$$

be a local quadratic approximation to $f(x+s)$, where B is a symmetric approximation to the current Hessian matrix. Then Powell's hybrid approach solves the problem

$$\min\{\Psi(s): s \in P_H, \|s\| \le \Delta\} \tag{5.4.2}$$

where P_H is the negative gradient direction if $\|s_G^*\| \le \Delta$; otherwise, P_H is the hybrid path

$$P_H := \{s: s = s_G^* + \theta(s_N - s_G^*); 0 \le \theta \le 1\} \tag{5.4.3}$$

The vector s_N is the quasi-Newton direction and s_G^* is the minimum along the negative gradient direction. Steihaug[1983] proposed *approximately* solving (5.4.2) by the linear conjugate gradient method in the following way. (For simplicity, we assume that the pre-conditioning matrix is the identity matrix.)

Algorithm For The Locally Constrained Problem

1. $s_0 \leftarrow 0$, $r_0 \leftarrow -\nabla f(x)$, $d_0 \leftarrow r_0$, $i \leftarrow 0$.
2. If $(d_i^T B d_i) > 0$ then go to 3.
   Else $s \leftarrow s_i + \tau d_i$, where $\|s\| = \Delta$, Exit(s)
3. $s_{i+1} \leftarrow s_i + \alpha_i d_i$, where $\alpha_i = \dfrac{r_i^T r_i}{d_i^T B d_i}$

   If $\|s_{i+1}\| < \Delta$ then go to 4.
   Else $s \leftarrow s_i + \tau d_i$, where $\|s\| = \Delta$, Exit(s)
4. $r_{i+1} \leftarrow r_i - \alpha_i B d_i$

   If $\dfrac{\|r_{i+1}\|}{\|\nabla f(x)\|} \le \eta$, then $s \leftarrow s_{i+1}$, Exit(s)
   Else
5. $\beta_i \leftarrow \dfrac{r_{i+1}^T r_{i+1}}{r_i^T r_i}$, $d_{i+1} \leftarrow r_{i+1} + \beta_i d_i$

   $i \leftarrow i+1$, go to 2.

Consider the piecewise linear path $P(i)$ that connects x to s_0, s_0 to $s_1, ..., s_{i-1}$ to s_i. It can be proven that $\Psi(s_j)$ is strictly decreasing for $j=0,...,i$; since α_j minimizes Ψ along d_j it follows that Ψ is strictly decreasing along $P(i)$. Moreover, $\|s_j\|$ is strictly increasing, for $j=0,...,i$. Therefore, if $\eta = 0$ then the algorithm solves the problem

$$\min\{\Psi(s): s \in P(i), \|s\| \le \Delta\} \tag{5.4.4}$$

and the minimum is unique. Note that, in exact arithmetic, s_{i-1} minimizes Ψ over $<d_0,...,d_{i-2}>$. If B is positive definite and $\eta = 0$ then the solution will approach the dogleg solution as i increases. In addition, if a direction of negative curvature is encountered then the solution will satisfy $\|s\| = \Delta$. The role of η is explained in Section 4.2.1.1 - in order to ensure superlinear convergence it is imperative that $\eta_k \to 0$.

This algorithm, which solves the local quadratic problem, can be imbedded in a trust region strategy, similar to that described in Section 4.4.1. Steihaug proved that, under reasonable circumstances, such an algorithm results in

$$\lim \inf \{\|\nabla f(x_k)\|\} = 0.$$

In addition, superlinear convergence can be proved under slightly stronger assumptions.

5.4.2 Nonlinear Conjugate Gradients

If $f(x)$ is a quadratic function with positive definite Hessian H, then the linear conjugate gradient method can be expressed as follows. Let g_k denote the gradient of f at x_k.

$$
\begin{aligned}
&k \leftarrow 1,\ x_1 \leftarrow 0,\ p_1 \leftarrow -g_1 \\
&\text{while } k \leq n \text{ do} \\
&\quad \alpha_k \leftarrow \frac{g_k^T g_k}{p_k^T H p_k} \\
&\quad x_{k+1} \leftarrow x_k + \alpha_k p_k \\
&\quad \beta_k \leftarrow \frac{g_{k+1}^T g_{k+1}}{g_k^T g_k}, \quad \text{if } \beta_k = 0, \quad \text{Exit} \\
&\quad p_{k+1} \leftarrow -g_{k+1} + \beta_k p_k \\
&\quad k \leftarrow k+1 \\
&\text{od}
\end{aligned}
$$

We can interpret α_k as the optimal step along direction p_k. Therefore, for an *arbitrary nonlinear* function f, it is reasonable to define $\alpha_k = \text{argmin} f(x_k + \alpha p_k)$. If the guard to the do-loop is suitable modified, say 'while $\|g_k\| > \epsilon do$', then the Fletcher-Reeves[1964] algorithm is defined. (It may be advisable/necessary to restart after a number of iterations.) The most attractive feature of this algorithm is its' storage requirement - only 3 vectors need be stored. If α_k is determined exactly, then it is easy to see that p_k is a descent direction:

$$g_k^T p_k = g_k^T [-g_k + \beta_{k-1} p_{k-1}] = -\|g_k\|^2 < 0$$

Unfortunately, it is usually not feasible to determine the exact minimizer of $f(x_k + \alpha p_k)$. As a result, it is possible that the descent direction property will be lost. There are a number of possible remedies including a restart and the imposition of certain line search conditions: Gill, Murray, and Wright[1981] discussed various possibilities. We will describe one possibility due to Shanno[1978].

Suppose that instead of defining p_{k+1} as in the algorithm above, we use

$$\bar{p}_{k+1} \leftarrow -[(I - \frac{p_k y_k^T}{p_k^T y_k})(I - \frac{y_k p_k^T}{p_k^T y_k}) + \frac{p_k s_k^T}{p_k^T y_k}] g_{k+1} \qquad (5.4.5)$$

where $s_k := x_{k+1} - x_k$. Inspection of (5.4.5) reveals that $[\cdot]$ defines the BFGS update (with respect to the inverse Hessian) where the 'current' Hessian approximation is the identity matrix. Therefore $\bar{p}_{k+1}$ must be a descent direction if $y_k^T p_k > 0$.

However, an exact line search implies $p_k^T g_{k+1} = 0$: under these circumstances it is easy to show that formula (5.4.5) generates a set of conjugate directions when applied to a quadratic function.

This equivalence allows for a nonlinear conjugate gradient algorithm without requiring exact line searches. In addition, it suggests that we can view conjugate gradients as a *memory-less* quasi-Newton method. (The relationship between BFGS and conjugate gradients is a close one. Nazareth[1979] demonstrated that the two methods are *identical* when applied to a strictly convex quadratic with exact line searches.)

Limited Memory Quasi-Newton Methods

A natural extension of the memory-less quasi-Newton method is a limited memory quasi-Newton method. Such a method computes p_k as $-M_k g_k$ where M_k is a positive definite matrix obtained by updating the identity matrix with a limited number of quasi-Newton corrections. Hence M_k will be an approximation to the inverse of the Hessian. Various implementations of these ideas have been suggested - see Gill, Murray and Wright [1981] for references.

5.4.3 The Pre-Conditioning Matrix

Both the linear and nonlinear conjugate gradient methods can be improved with the use of suitable pre-conditioning matrices. There is a discussion of related ideas in Section 1.3.1.2 - this discussion is particularily relevant to the linear conjugate gradient method. A possibility not mentioned there is to use a quasi-Newton approximation to the Hessian inverse as a pre-conditioner. For example, the limited memory quasi-Newton method discussed above can be viewed as a nonlinear conjugate gradient method with a changing pre-conditioner. Other possibilities include the use of the BFGS update to define a diagonal preconditioner - see Gill, Murray and Wright[1981].

5.5 Global Convergence and Direct Equation Solvers

Most of the sparse Hessian approximation techniques discussed in this section allow for an *indefinite* approximation B. In Section 5.4 we discussed a technique which yields global convergence properties and which uses an iterative approach to the quasi-Newton equation. It is desirable to also have available global techniques which employ direct sparse linear equation solvers. Here we will discuss two trust region possibilities. We assume that B is nonsingular.

5.5.1 Powell's[1970] Dogleg Algorithm

This algorithm is a direct adaptation of the nonlinear least squares method discussed in Section 4.4.4. The major difference is that here we must allow for indefinite matrices B.

A trial correction to x, $x^+ \leftarrow x + s$, is obtained by approximately solving the local quadratic model

$$\min\{\Psi(s) \colon \|s\| \leq \Delta\} \tag{5.5.1}$$

where $\Psi(s) = g^T s + \dfrac{1}{2} s^T B s$ and g denotes $\nabla f(x)$. An approximate solution to (5.5.1) is obtained by solving

$$\min\{\Psi(s) \colon s \,\epsilon\, P_H, \|s\| \leq \Delta\} \tag{5.5.2a}$$

where $P_H := -g$ if $\|s_G^\circ\| \geq \Delta$
Otherwise, $\{ \; g^T Bg > 0 \; ; \; \|s_G^\circ\| < \Delta \; \}$ and

$$P_H := \{s \colon s = s_G^\circ + \theta(s_N - s_G^\circ), 0 \leq \theta \leq 1\} \tag{5.5.2b}$$

The Newton step and the optimal gradient step are denoted by s_N and s_G° respectively. This simple algorithm can be described as follows.

Dogleg Algorithm
1. Compute $\sigma = g^T Bg$
   If $\sigma \leq 0$ then $s \leftarrow -\alpha g$ where $\|s\| = \Delta$, $\alpha > 0$, Exit(s)
   Else
2. If $\dfrac{\|g\|^3}{g^T Bg} \geq \Delta$
   then $s \leftarrow -\alpha g$ where $\|s\| = \Delta$, $\alpha > 0$, Exit(s)
   Else $\alpha \leftarrow \dfrac{\|g\|^2}{g^T Bg}$, $s_G^\circ \leftarrow -\alpha g$
3. Compute $s_N \leftarrow -B^{-1} g$
   $s \leftarrow s_G^\circ + \beta(s_N - s_G^\circ)$,
   where s solves (5.5.2a) with P_H defined by (5.5.2b).

Note that if Δ is small then the correction s will tend toward the negative gradient direction. On the other hand, if Δ is sufficiently large and B is positive definite then s will be the Newton step. This local algorithm can be imbedded in a Δ-modification strategy similar to that described in Section 4.4.1.

5.5.2 Solving The Quadratic Model

In this section we consider another approach to the solution of problem (5.5.1). This description is based on More' and Sorensen[1981,1982]. Lemma 3.5 of More' and Sorensen[1982] completely characterizes the solution to problem (5.5.1):

Lemma 5.5.1 *Let* $\Psi := g^T s + \dfrac{1}{2} s^T B s$ *and let* $\Delta > 0$ *be given. The point* s_* *solves the problem*

$$\min\{\Psi(s): \|s\| \leq \Delta\} \tag{5.5.3}$$

if and only if there is a $\lambda \geq 0$ *such that*

$$(B + \lambda I)s_* = -g, \quad \lambda(\Delta - \|s_*\|) = 0 \tag{5.5.4}$$

and $B + \lambda I$ *is positive semi-definite.*

Proof: Suppose that s_* and λ satisfy (5.5.4) with $B + \lambda I$ positive semi-definite. Clearly s_* minimizes the unconstrained quadratic function

$$\hat{\Psi}(s) = g^T s + \frac{1}{2} s^T (B + \lambda I)s \tag{5.5.5}$$

Hence, for any s, $\hat{\Psi}(s_*) \leq \hat{\Psi}(s)$, which implies

$$g^T s_* + \frac{1}{2} s_*^T B s_* \leq g^T s + \frac{1}{2} s^T B s + \frac{1}{2} \lambda(\|s\|^2 - \|s_*\|^2) \tag{5.5.6}$$

for all s. But $\lambda \|s_*\|^2 = \lambda \Delta^2$, $\lambda \geq 0$ and since $\|s\|^2 \leq \Delta^2$ it follows that s_* solves (5.5.3).

Suppose s_* solves (5.5.3). Clearly if $\|s_*\| < \Delta$ then s_* is an unconstrained minimizer and (5.5.4) follows with B positive semi-definite. Otherwise, $\|s_*\| = \Delta$ and s_* solves

$$\min\{\Psi(s): \|s\| = \Delta\}$$

and therefore, using first-order necessary conditions for constrained optimization, (5.5.4) holds for some $\lambda \geq 0$. It only remains to show that $B + \lambda I$ is positive semi-definite. But for every s with $\|s\| = \|s_*\| = \Delta$, we have

$$g^T s + \frac{1}{2} s^T B s \geq g^T s_* + \frac{1}{2} s_*^T B s_*$$

or, substituting $-(B + \lambda I)s_*$ for g,

$$\tfrac{1}{2}(s - s_*)^T (B + \lambda I)(s - s_*) \geq 0$$

Hence $B + \lambda I$ is positive semi-definite and the theorem is established.

In order to analyze the systems (5.5.4), consider the eigenvalue/eigenvector decomposition

$$B = V \Lambda V^T, \quad V^T V = 0, \quad \Lambda = diag(\lambda_1, ..., \lambda_n), \quad with \ \lambda_1 \geq \lambda_2 \geq , \dots, \geq \lambda_n$$

Then $B + \lambda I = V[\Lambda + \lambda I]V^T$ and if α is defined by $-g = V\alpha$ then the system of equations in (5.5.4) becomes

$$(\Lambda + \lambda I)(V^T s) = \alpha$$

Therefore, for $\lambda > -\lambda_n$,

$$s = \sum_{i=1}^{n} (\frac{\alpha_i}{\lambda_i + \lambda}) v_i \tag{5.5.7}$$

There are three basic cases to consider.

Case 1: $\lambda_n > 0$.

If $\|B^{-1}g\| \leq \Delta$ then the solution is $s_* = -B^{-1}g$. If $\|B^{-1}g\| > \Delta$ then, from (5.5.7) it is clear that $\|s\|$ decreases to zero as λ increases from $-\lambda_n$. Hence for some $\lambda > -\lambda_n$, $\|s(\lambda)\| = \Delta$.

Case 2: $\lambda_n \leq 0$ and $\alpha_i \neq 0$ for some $i \epsilon I := \{i : \lambda_i = \lambda_n\}$.

Again, from (5.5.7), $\|s(\lambda)\|$ decreases to zero as λ increases from $-\lambda_n$; $\|s(\lambda)\|$ increases to infinity as λ approaches $-\lambda_n$ from above. Therefore for some $\lambda \epsilon (-\lambda_n, \infty)$, $\|s(\lambda)\| = \Delta$.

Case 3: $\lambda_n \leq 0$ and $\alpha_i = 0$ for all $i \epsilon I$.

We can dispose of one subcase immediately. Suppose that for some $\lambda > -\lambda_n$, $\|s(\lambda)\| > \Delta$. From (5.5.7), $\|s(\lambda)\|$ decreases to zero as λ increases to infinity and hence $\|s(\lambda)\| = \Delta$ has a solution. We are left with the one possibility which More' and Sorensen[1981] label *the hard case*. Specifically, $\{\lambda_n \leq 0$ and $\alpha_i = 0$, for *all* $i \epsilon I\}$ and for all $\lambda > -\lambda_n$, $\|s(\lambda)\| < \Delta$. Let $\bar{s} = \sum_{i \epsilon \bar{I}} [(\frac{\alpha_i}{\lambda_i - \lambda_n}) v_i]$, where $\bar{I}$ is the complement of I. If $\|\bar{s}\| = \Delta$ then $s_* \leftarrow \bar{s}$ and we are done. Otherwise, let

$$s_* = \bar{s} + \sum_{i \epsilon I} \mu_i v_i$$

such that $\|s_*\| = \Delta$.

Computing An Approximate Solution

In this section we will sketch some of the ideas behind the approach of Moré and Sorensen [1981] (see also Gay[1981]) to obtain an approximate solution to (5.5.3). It is important to bear in mind that the computational effort must be suitable for use within an optimization algorithm. That is, it is expected that many locally constrained problems must be solved and therefore, to remain practical, the computation of a solution to (5.5.3) cannot be unduly expensive.

Let $s(\lambda)$ be defined by

$$s(\lambda) = -(B + \lambda I)^{-1}g$$

for $\lambda \geq 0$, *in* $(-\lambda_n, \infty)$. If (5.5.3) has a solution on the boundary, and we are not dealing with the hard case, then we seek a solution to

$$\|s(\lambda)\| = \Delta$$

But, from (5.5.7), $s(\lambda)$ has a pole at $\lambda = -\lambda_n$: Newton's method will converge extemely slowly when λ is near $-\lambda_n$. It is preferable to find a zero of the reciprocal function

$$\tilde{\Psi}(\lambda) := \frac{1}{\Delta} - \frac{1}{\|s(\lambda)\|}$$

which is almost linear for $\lambda > -\lambda_n$. Newton's method can now be applied in a straightforward way when the optimal $\lambda \in (-\lambda_n, \infty)$. Note that every evaluation will involve the solution of a linear system; however, since $\tilde{\Psi}$ is almost linear in the interval of interest we can expect extremely rapid convergence.

In the hard case, it is necessary to compute an approximate eigenvector of B corresponding to λ_n. Moré and Sorensen[1981] determine a suitable approximation z by using the LINPACK condition number estimation technique. This involves determining z such that $\|Rz\|$ is as small as possible, where R^T is the Cholesky factor of a positive definite matrix $B + \bar{\lambda}I$ and $\|z\| = 1$. The interested reader should consult Cline, Moler, Stewart, and Wilkinson [1979] and Moré and Sorensen[1981] for more details.

Computationally, the method of Moré and Sorensen seems quite satisfactory: On a set of test problems, the average number of factorizations per iteration was 1.6. At this point it is not clear that such a technique is suitable for large-scale problems (assuming the factors can be held in main memory). In particular, the dogleg algorithm can be implemented to average less than 1 factorization per iteration. On the other hand, the approximate solution yielded by the dogleg approach is more restricted and therefore it is expected that the number of iterations required by the dogleg algorithm will be greater.

5.5.6 Convergence Results For Trust Region Methods

Trust region methods are appealing for large sparse problems because they handle indefinite matrices in a geometrically meaningful way. Moreover, strong convergence results can be proven under fairly mild assumptions. In this section we will review some of the convergence results. In particular we will present some of the results given in More' [1982], though our presentation is *not quite as general*. For example, we will drop the scaling matrices for simplicity of presentation. The interested reader is encouraged to study the paper of More' to appreciate the full generality of the results and enjoy their proofs. The paper of Powell[1970] is also important reading since Powell provided a first version of many of the theorems. (See also Fletcher[1980] and Sorensen[1982].)

Firstly we present the general trust region algorithm as described by More'[1982].

Let x_0, Δ_0 be given.
For $k = 0,1,...$
    1. Compute $f(x_k)$ and Ψ_k.
    2. Determine an approximate solution s_k to (5.5.1).
    3. $\rho_k \leftarrow \dfrac{f(x_k + s_k) - f(x_k)}{\Psi_k(s_k)}$
    4. If $\rho_k > \mu$ then $x_{k+1} \leftarrow x_k + s_k$, otherwise $x_{k+1} \leftarrow x_k$.
    5. Update Δ_k: Given $0 < \mu < \eta < 1$, $\gamma_1 < 1 < \gamma_2$
        a. If $\rho_k \leq \mu$ then $\Delta_{k+1} \in (0, \gamma_1 \Delta_k]$
        b. If $\rho_k \in (\mu, \eta)$ then $\Delta_{k+1} \in [\gamma_1 \Delta_k, \Delta_k]$
        c. If $\rho_k \geq \eta$ then $\Delta_{k+1} \in [\Delta_k, \gamma_2 \Delta_k]$.

The first theorem is applicable to both algorithms presented earlier in this section. Actually, it is applicable to any trust region method which fits into the mode described above and which is gradient-related:

$$\Psi_k(s_k) \leq \beta_1 \min\{\Psi_k(s): s = -\alpha g_k, \|s_k\| \leq \Delta_k\} \tag{5.5.8}$$

The following lemma, due to Powell[1970], is needed to establish Theorem 5.5.3 and follows immediately from (5.5.8).

Lemma 5.5.2. $-\Psi(s_k) \geq \dfrac{1}{2} \beta_1 \|g_k\| \min\{\Delta_k, \dfrac{\|g_k\|}{\|B_k\|}\}$.

We will assume that there exists $\sigma_1 > 0$ such that $\|B_k\| \leq \sigma_1$. The following result (Powell[1970]) states that $\{\|g_k\|\}$ is not bounded away from zero.

Theorem 5.5.3. If f is continuously differentiable and bounded below and (5.5.8) is satisfied then

$$\lim_{k \to \infty} \inf \|g_k\| = 0$$

It is desirable to replace " lim inf " with " lim". More', building on work of Thomas[1975], showed that the sequence is convergent under slightly stronger assumptions. A critical ingredient of this result is that μ (in step 4. of the algorithm) is nonzero.

Theorem 5.5.4. Under the assumptions of Theorem 5.5.3 and further assuming that ∇f is uniformly continuous, then

$$\lim_{k \to \infty} \|g_k\| = 0$$

While the above results are very positive and suggest probable convergence of $\{x_k\}$ in practise, they do not ensure it. Therefore we are left with 2 important questions: when does the sequence $\{x_k\}$ converge and when does it do so superlinearly ? We can expect superlinear convergence (given convergence) only if the sequence $\{\Delta_k\}$ is bounded away from zero. Condition (5.5.8) will produce the desired effect under stronger assumptions on f provide we assume that $B_k = \nabla^2 f(x_k)$. The following theorem follows Theorem 4.19 of More[1982].

Theorem 5.5.5 Let f be twice continuously differentiable and bounded below and assume that $\nabla^2 f$ is bounded on the level set $\{x \in R^n : f(x) \leq f(x_0)\}$ Further assume that (5.5.8) is satisfied with $B_k = \nabla^2 f(x_k)$. If x_* is a limit point of $\{x_k\}$ with $\nabla^2 f(x_*)$ positive definite, then the sequence $\{x_k\}$ converges to x_*, all iterations are eventually successful, and $\{\Delta_k\}$ is bounded away from zero.

Proof: We first prove that $\{x_k\}$ converges to x_*. Choose δ so that $\nabla^2 f(x)$ is positive definite for $\|x - x_*\| \leq \delta$. Since $\Psi_k(s_k) \leq 0$, there is an $\epsilon_1 > 0$ such that

$$\|x_k - x_*\| \leq \delta \Longrightarrow \epsilon_1 \|s_k\| \leq \|\nabla f(x_k)\| \qquad (5.5.9)$$

Theorem 5.5.4 guarantees that $\{\nabla f(x_k)\}$ converges to zero; hence, there is a $k_1 > 0$ such that

$$\|\nabla f(x_k)\| \leq \frac{1}{2} \epsilon_1 \delta, \ k \geq k_1.$$

Therefore, by (5.5.9), if $\|x_k - x_*\| \leq \frac{1}{2} \delta$ for $k \geq k_1$, then $\|x_{k+1} - x_*\| \leq \delta$. It is now easy to see, using the local convexity of f around x_* and the fact that x_* is a limit point], there is an index $k_2 \geq k_1$ with $\|x_{k_2} - x_*\| \leq \frac{1}{2} \delta$ such that

$$f(x) \leq f(x_k), \|x - x_*\| \leq \delta \Longrightarrow \|x - x_*\| < \frac{1}{2} \delta$$

Hence $\|x_k - x_*\| \leq \frac{1}{2} \delta$ for $k \geq k_2$. This proves that $\{x_k\}$ converges to x_*.

We now show that all iterations are successful. (An iteration is *successful* if $\rho_k > \mu$.) Since $\|s_k\| \leq \Delta_k$ and $\epsilon_1 \|s_k\| \leq \|\nabla f(x_k)\|$, it follows that, using Lemma 5.5.2,

$$-\Psi_k(s_k) \geq \epsilon_2 \|s_k\|^2, \tag{5.5.10}$$

for some $\epsilon_2 > 0$. But clearly,

$$|f(x_k + s_k) - \Psi(s_k)| \leq \frac{1}{2} \|s_k\|^2 \max\{\|\nabla^2 f(x_k + \tau s_k) - \nabla^2 f(x_k)\|: 0 \leq \tau \leq 1\}$$

which implies, using (5.5.10), that $|\rho_k - 1|$ converges to zero, and the proof is complete.

This theorem does not establish convergence to a local minimizer since $\{x_k\}$ may not have a limit point; if it does, it may not satisfy second order sufficiency conditions. Nevertheless, it suggests that, in practise, convergence will usually be observed with $\{\Delta_k\}$ bounded away from zero.

It is to be expected that an algorithm which (approximately) solves (5.5.3) would possess stronger convergence properties. This expectation is realized. Let $\{x_k\}$ be generated by an algorithm, in the trust region mode, which satisfies

$$\Psi_k(s_k) \leq \beta_2 \min\{\Psi_k(s): \|s\| \leq \Delta\} \tag{5.5.11}$$

where

$$\beta_2 > 0, \quad \text{and} \quad \Psi_k(s) = g_k^T s + \frac{1}{2} s^T \nabla^2 f(x_k) s$$

(The algorithm discussed in Section 5.5.2 satisfies this condition.) The following theorem is given in More[1982] and to a large extent is a summary of results given in Fletcher[1980], Sorensen[1982], and More and Sorensen[1981].

Theorem 5.5.6 Let the assumptions of Theorem 5.5.5 hold. In addition, assume that (5.5.11) is satisfied. Then,

a) $\{\nabla f(x_k)\}$ converges to zero.

b) If $\{x_k\}$ is bounded then there is a limit point x_*, with $\nabla^2 f(x_*)$ positive semi-definite.

c) If x_* is an isolated limit point of $\{x_k\}$ then $\nabla^2 f(x_*)$ is positive semi-definite.

d) If $\nabla^2 f(x_*)$ is nonsingular for some limit point, x_*, of $\{x_k\}$ then $\nabla^2 f(x_*)$ is positive definite, $\{x_k\}$ converges to x_* and all iterations are eventually successful.

A trust region algorithm will exhibit superlinear convergence if $\{x_k\}$ converges to a point satisfying second-order sufficiency conditions and there exists $\beta_3 > 0$ such that if $\|s_k\| < \beta_3 \Delta_k$ then

$$\|\nabla f(x_k) + \nabla^2 f(x_k) s_k\| \leq \epsilon_k \|\nabla f(x_k)\|$$

where ϵ_k converges to zero. Powell's[1970] dogleg algorithm and the algorithm of Section 5.5.2 both satisfy these conditions with $\beta_3 = 1$ and $\beta_3 = 1 - \sigma$, respectively. The conjugate gradient algorithm of Steihaug (Section 5.4.1) also satisfies these conditions with $\beta_3 = 1$.

5.6 Exercises

1. Complete the following argument. If it is possible to obtain a minimum symmetric coloring of an arbitrary *bipartite* graph in polynomial time, then it is also possible to obtain a minimum (ordinary) coloring of an arbitrary (general) graph in polynomial time. To see this, let $G = (V,E)$ be an arbitrary graph and construct a bipartite graph as follows. Let $v_1, \ldots, v_n$ be the vertices of G, and let $e_1,\ldots,e_m$ be the edges of G. For each edge $e_l = (v_i,v_j)$ define a bipartite graph B_l with vertices

$$\{ v_i, v_j, w_1^{(l)}, \ldots, w_n^{(l)} \}$$

and edges

$$(v_i, w_k^{(l)}), \quad (v_j, w_k^{(l)}), \quad k = 1,\ldots,n .$$

Now define a bipartite graph B by setting

$$V(B) = V(G) \bigcup \{ w_k^{(l)} : 1 \le k \le n, \ 1 \le l \le m \}$$

and

$$E(B) = \bigcup_{l=1}^{m} E(B_l) .$$

Now, ...

2. Verify that $z^T R_{ut} z > 0$, in the proof of Theorem 5.2.1.

3. Verify that the method of Shanno and Toint (Section 5.2) can be implemented without ever computing quantities which are not in $S(B)$.

4. Verify statements (5.3.4), (5.3.5), and (5.3.7). In addition, show that positive semi-definiteness is inherited by both DFP and BFGS updates, provided $s_i^T y_i > 0$.

5. Show that statement (5.4.5) yields the conjugate gradient method when exact line searches are performed and the problem is quadratic.

6. Is it feasible to expect that problem (5.5.1) can be solved, globally, by an efficient algorithm when $\|\cdot\|$ denotes the l_∞ (or l_1) norm ? Why or why not ?

5.7 Research Problems

1. Investigate theoretical and computational properties of a trust region method based on $\|\cdot\|_\infty$.

2. Investigate an approach to the large sparse unconstrained optimization

problem which allows for storage of the sparse Cholesky factor of H_k and several n-vectors representing recent BFGS updates.

5.8 References

Cline, A.K., Moler, C.B., Stewart, G.W., and Wilkinson, J.H. [1979]. *An estimate of the condition number of a matrix*, SIAM J. Numer. Anal. 16, 368-375.

Coleman, T.F and Moré [1982]. *Estimation of sparse Hessian matrices and graph coloring problems*, Technical Report 82-535, Department of Computer Science, Cornell University, Ithaca, NY.

Dembo, R., Eisenstat, S., and Steihaug [1982]. *Inexact Newton Methods*, SIAM Journal on Numerical Analysis 19, 400-408.

Dennis, J.E. and Moré [1977]. *Quasi-Newton methods, motivation and theory*, SIAM Review 19, 46-89.

Dennis, J.E. and Schnabel, R.B. [1979]. *Least change secant updates for quasi-Newton methods*, SIAM Review 21, 443-459.

Dennis, J.E. and Schnabel, R.B.[1983]. *Numerical Methods for Unconstrained Optimization and Nonlinear Equations*, Prentice-Hall.

Fletcher,R. [1980]. *Practical Methods of Optimization: Unconstrained Optimization*, John Wiley and Sons.

Fletcher, R. and Reeves, C.M. [1964]. *Function minimization by conjugate gradients*, Computer Journal 6, 163-168.

Gay, D.M. [1981]. *Computing optimal locally constrained steps*, SIAM J. Sci. Stat. Comput. 2, 186-197.

Griewank, A. and Toint Ph. L [1982]. *Partitioned Variable Metric Updates for Large Structured Optimization Problems*, Numerische Mathematik 39, 429- 448.

Griewank, A. and Toint, Ph. L [1983]. *On the existence of convex decompositions of partially separable functions*, to appear in Mathematical Programming.

Gill, P.E., Murray, W., and Wright, M.[1981]. *Practical Optimization*, Academic Press, New York.

Gill, P.E., Murray, W., Saunders, M.A., and Wright, M.H. [1983]. *Computing Forward-Difference Intervals for Numerical Optimization*, SIAM Journal on Scientific and Statistical Computing 4, 310-321.

Marwil, E. [1978]. *Exploiting sparsity in Newton-type methods*, PhD. Dissertation, Applied Mathematics, Cornell University, Ithaca, NY.

Moré, J.J, [1982]. *Recent developments in algorithms and software for trust region methods*, Technical Report ANL/MCS-TM-2, Argonne National Laboratory, Argonne, Il.

Moré, J.J. and Sorensen, D.C.[1982]. *Newton's Method* , Technical Report ANL-82-8, Argonne National Laboratory, Argonne, Il.

More', J.J. and Sorensen, D.C. [1981]. *Computing a Trust Region Step* , Technical Report ANL-81-83, Argonne National Laboratory, Argonne, Il.

Nazareth, L.[1979]. *A relationship between the BFGS and conjugate gradient algorithms and its implication for new algorithms*, SIAM J. on Numerical Analysis, 16, 794-800.

O'Leary, D.P. [1980]. *A discrete Newton algorithm for minimizing a function of many variables*, Technical Report 910, Computer Science Center, University of Maryland, College Park, Maryland.

Paige, C. and Saunders, M. [1982]. *LSQE: An algorithm for sparse linear equations and sparse least squares*, ACM Trans. on Math. Software, 8, 43-71.

Powell, M.J.D. [1970]. *A new algorithm for unconstrained optimization*, in 'Nonlinear Programming', J.B. Rosen, O.L. Mangasarian, and K.Ritter, eds., Academic Press, 31-66.

Powell, M.J.D. and Toint, Ph. L.[1979]. *On the estimation of sparse Hessian matrices*, SIAM J. on Numerical Analysis 16, 1060-1074.

Powell, M.J.D. and Toint, Ph. L [1981]. *The Shanno-Toint procedure for updating sparse symmetric matrices*, IMA Journal of Numerical Analysis 1, 403-413.

Powell, M.J.D. [1981]. *A note on quasi-Newton formulae for sparse second derivative matrices*, Math. Prog. 20, 144-151.

Shanno, D.F. [1978]. *Conjugate-gradient methods with inexact searches*, Math. of Oper. Res. 3, 244-256.

Shanno, D.F [1980]. *On the variable metric methods for sparse Hessians*, Math. Comp. 34, 499-514.

Sorensen, D.C. [1981]. *An example concerning quasi-Newton estimation of a sparse Hessian*, SIGNUM Newsletter, 16, 8-10.

Sorensen, D.C [1982]. *Trust region methods for uncontrained optimization*, SIAM J. Numer. Anal. 19, 409-426.

Steihaug,T [1983]. *The conjugate gradient method and trust regions in large scale optimization*, SIAM Journal on Numerical Analysis 20, 626-637.

Thomas, S.W. [1975]. *Sequential estimation techniques for quasi-Newton algorithms*, Ph.D. Dissertation, Cornell University, Ithaca, NY.

Toint, Ph. L [1977]. *On sparse and symmetric updating subject to a linear equation*, Math. Comp. 32, 839-851.

Toint, Ph.L [1981a]. *A sparse quasi-Newton update derived variationally with a non-diagonally weighted Frobenius norm*, Math. Comp. 37, 425-434.

Toint, Ph.L [1981]. *A note on sparsity exploiting quasi-Newton methods*, Mathematical Programming 21, 172-181.

Thapa, M.N. [1979]. *A note on sparse quasi-Newton methods*, Technical Report 79-13, Dept. of Operations Research, Stanford University, Stanford, CA.

Chapter 6   Large Sparse Quadratic Programs

We will consider only certain aspects of this important problem here. The literature on this topic is rich with diverse algorithmic ideas. In these notes we will discuss some computational issues with respect to a particular problem formulation. Fletcher[1981], Gill, Murray, and Wright[1981] and Gill, Murray, and Wright[1983] provide additional detail and references.

6.1 Large Quadratic Programming Problems with Equality Constraints

In this section we consider problems of the form

$$\min\{q(x),\ Ax = b\} \tag{6.1.1}$$

where q is a quadratic function:

$$q(x) := g^T x + \frac{1}{2} x^T Bx,$$

and B is symmetric. The matrix A is t by n, $t \leq n$ and of rank t. If x_* is a solution then there exists λ_* such that

$$\begin{bmatrix} B & A^T \\ A & 0 \end{bmatrix}\begin{bmatrix} x_* \\ \lambda_* \end{bmatrix} = \begin{pmatrix} -g \\ b \end{pmatrix} \tag{6.1.2}$$

Of course if (x_*, λ_*) solves (6.1.2), and x_* is feasible, it does not follow that x_* solves (6.1.1) since (6.1.2) merely ensures that x_* is a stationary point. It is true, however, that if x_* satisfies (6.1.2) then either x_* solves (6.1.1) or the problem is unbounded below.

There are at least 2 ways of solving (6.1.1) via (6.1.2). Firstly, it is possible to treat system (6.1.2) as a large sparse symmetric system of equations and to invoke a sparse solver, iterative or direct. Note that the matrix is *not* positive definite; indeed, B need not be. (The matrix in (6.1.2) is nonsingular however, provided a unique solution to (6.1.1) exists.) Secondly, if it is known that B is positive definite then advantage can be taken of the block structure of (6.1.2).

6.1.1 Range Space Methods

This approach is applicable when it is known that B is positive definite. The linear system (6.1.2) can then be solved in a manner similar to the block method discussed in Section 1.1.4. In particular, a useful asymmetric factorization is

$$\begin{bmatrix} B & A^T \\ A & 0 \end{bmatrix} = \begin{pmatrix} B & 0 \\ A & C \end{pmatrix}\begin{bmatrix} I & B^{-1}A^T \\ 0 & I \end{bmatrix} \tag{6.1.3}$$

where $C := -AB^{-1}A^T$. The factorization can be computed in a way which is particularily attractive when t is fairly small.

Factor Step
  1) Factor $B = LL^T$
  2) For $i=1,...,t$: Solve $By = a_i$, $c_i \leftarrow -Ay$
  3) Factor $-C = \bar{L}\bar{L}^T$

Column i of C is defined by step 2, where a_i is column i of A_i^T. Note that if t is small then C can be treated as a dense matrix. Storage is needed for the sparse factor L, the small dense factor $\bar{L}$, and the sparse matrix A.

The solve step utilizes the factorization(6.1.3):

Solve Step
  1) Solve $Bt = -g$, $\bar{b} \leftarrow b-At$
  2) Solve $C\lambda = \bar{b}$, $\bar{g} \leftarrow -g-A^T\lambda$
  3) Solve $Bx = \bar{g}$

There are other ways to exploit the asymmetric factorization (6.1.3). For example, if it is possible to store an n by t dense matrix, then a procedure involving $A^T = QR$ can be followed as described by Gill, Murray, and Wright [1981]. This approach is probably preferable from a numerical point of view since then the matrix ' C ' is $Q^TB^{-1}Q$ where Q is orthogonal. Hence the conditioning of C is not worse than the conditioning of B.

If B is not positive definite then the methods mentioned above cannot be applied. However, the solution to (6.1.1) is unaffected if B is replaced with $B + \omega A^TA$; if ω is sufficiently large then $B + \omega A^TA$ will be positive definite. There are two obvious difficulties. Firstly, the task of determining a suitable value for ω is not trivial. Secondly, the density of the matrix $B + \omega A^TA$ is greater than the density of B.

A range space method becomes less attractive as t increases. On the other hand, a *null space method* becomes more attractive as t increases. In addition, a null space method does not depend on the positive definiteness of B.

6.1.2 Null Space Methods

Let Z be an n by $n-t$ matrix with columns that form an orthonormal basis for the null space of A. Then every feasible point can be expressed as $x = Zy + \bar{x}$, where $\bar{x}$ is an arbitrary feasible point. Therefore, problem (6.1.1) can be written as the unconstrained problem

$$\min\{y^Tw + \tfrac{1}{2}\; y^T(Z^TBZ)y\} \tag{6.1.4}$$

where $w := Z^Tg + Z^TB\bar{x}$. A solution to (6.1.4) satisfies

$$(Z^TBZ)y = -w \tag{6.1.5}$$

with $Z^T BZ$ positive semi-definite (if $Z^T BZ$ is indefinite then (6.1.1) is unbounded below). If $Z^T BZ$ is positive definite, then the solution is unique.

Orthogonal Approaches

If it is feasible to store $n-t$ dense n-vectors, then it may be possible to use an orthonormal basis for the null space of A and solve (6.1.5) directly. For example, a feasible point can be found by solving

$$AA^T \hat{x} = b, \quad \bar{x} \leftarrow A^T \hat{x} \tag{6.1.6}$$

This can be achieved, without storing Q, by applying a sequence of Givens transformations:

$$Q^T A^T = Q_m^T \cdots Q_1^T A^T = \binom{R}{0}$$

A solution to (6.1.6) is realized by solving

$$R^T \tilde{x} = b, \quad R\hat{x} = \tilde{x}$$

Clearly, the solution to (6.1.5) presents no problem under the assumptions that $n-t$ is small and the projected Hessian, $Z^T BZ$, is positive definite. One difficulty remains however: How do we obtain Z without requiring excessive intermediate storage ? If this is not possible then it is important that Q be represented compactly. Recent work by George and Ng[1983] suggests that this *might* be possible, in general, however this question needs more work. In addition, if it is necessary to store Q (at least temporarily) then one needs to question the 'need' for storing Z explicitly, since all quantities can be computed using Q.

An interesting research question is this: Given that A is sparse (or t is small) is it possible to represent Z, compactly? In addition, is it possible to compute Z, efficiently, using limited space?   A satisfactory solution might be that Q is stored as a sequence of Givens transformations:

$$Q^T = Q_m^T \cdots Q_1^T$$

where m is acceptably small. (Note that the integer m depends, in part, on the order in which A is decomposed.) The product $Z^T v$ is then computed as $Z^T = [0 : I_{n-t}]Q^T v$.

When $n-t$ is not small then, in order to solve (6.1.5) directly, it is necessary that $Z^T BZ$ be sparse - it seems highly ambitious to achieve sparseness in the projected Hessian as well as orthonormality and 'sparseness' in Z. Therefore, such an approach seems restricted to an iterative solution of the positive definite system (6.1.5).

Elementary Transformations

As indicated in Chapter 5, question 2.7, the reduction of A^T to upper triangular form via Gauss transformations (and interchanges) yield a basis, Z, for $null(A)$. In particular, if G represents the product of Gauss transformations and interchanges (ie. $G = L^{-1}$) such that

$$GA^T = \binom{U}{0}$$

then we can define Z:

$$Z = G^T \begin{bmatrix} 0 \\ I_{n-t} \end{bmatrix}$$

Therefore, if G is stored as a sequence of Gauss transformations

$$G = G_m \cdots \cdots G_1$$

and m is suitably small, then Z is implicitly available. Again, it is unlikely that $Z^T BZ$ will be sparse - an iterative solution of (6.1.5) is called for when $(n-t)$ is not small.

Elimination Method

This approach is similar to that described in Section 2.6.2 and the linear programming basis partition approaches (Chapter 3). Let A be partitioned $A = (A_1, A_2)$ where A_1 is invertible. Then we can define Z as

$$Z = \begin{bmatrix} -A_1^{-1} A_2 \\ I \end{bmatrix}$$

The partition of A can be chosen so that the factors of A_1, say $A_1 = LU$, are sparse (with an eye to a small condition number also). Again, this approach seems to restrict the method of solution of (6.1.5) to an iterative scheme, unless $n-t$ is small.

6.2 Large QP's with Equality and Inequality Constraints

The general QP problem can be expressed as

$$\min\{q(x): Ax = b; l \leq x \leq u\} \tag{6.2.1}$$

This formulation seems most suitable for large scale optimization and it is consistent with most large scale LP formulations. In this section we will not discuss complete algorithms for the solution of (6.2.1). Rather, we will focus on some of the basic issues underlying any active set algorithm implemented with respect to formulation (6.2.1). An active set algorithm generates a sequence of feasible points. At each point a subset of the active set, called the *working set*, is defined. This set will define a local *equality* constrained problem in the folowing way. Suppose at x_k the variables 1,2,...,l are currently at their bounds and it has been *decided* to to keep these constraints active. That is, the new point, x_{k+1}, will be defined by $x_{k+1} \leftarrow x_k + \bar{p}_k$ and $\bar{p}_{k_i} = 0$, $i=1,...,l$. The following reduced equality constrained QP defines p_k, where $\bar{p}_k = (0, p_k)$:

$$\min\{g_k^T p_k + \frac{1}{2} p_k^T B_k p_k : A_k p_k = 0\} \tag{6.2.2}$$

where g_k consists of the last $n-l$ components of $\nabla q(x_k)$, B_k is the lower right-hand submatrix of B_k, and A_k consists of the last $n-l$ columns of A. The variables that are currently held at their bounds are called *bound* variables while the others are called *free* variables. Note that the number of free variable is always greater than t.

A solution to (6.2.2) can be computed using one of the techniques described in Section 6.1. Of course the computed solution may not be feasible in which case a restricted step must be taken. At the new point, x_{k+1}, one of two possibilites will generally occur. A new constraint may be added to the working set - that is, a free variable will become a bound variable. Alternatively, a constraint may be dropped from the working set - that is, a bound variable may become free. Hence B_{k+1} will differ from B_k in one row and column, and A_{k+1} will differ from A_k in one column.

We are led to a major computational concern. It is important that the matrix factorizations used in the solution of subproblem (6.2.2) be easily updated. It would be highly inefficient to solve each problem from scratch.

Matrix Modifications for Null Space Methods

The elementary transformation approach appears to have a serious difficulty: it is not clear how to update when columns of A are dropped. The orthogonal approaches have possibilities however. We will first describe a scheme, due to Gill, Golub, Murray, and Saunders[1974] which assumes that Q is explicitly available. We will then present a modification which needs only Z. Our development follows Lawson and Hanson[1974].

Suppose $Q^T A^T = \binom{R}{0}$, where A is t by m and $t < m$. Without loss of generality we will assume that the last column of A is to be removed. Note: We will drop the subscript k but it is important to remember that A refers to a selection of columns of the original constraint matrix. That is $A = [\bar{A} : v]$ and we seek $\bar{Q}, \bar{R}$ such that

$$\bar{Q}^T \bar{A}^T = \begin{bmatrix} \bar{R} \\ 0 \end{bmatrix}$$

Partition Q^T so that

$$\begin{bmatrix} Q_1 & s \\ u^T & \alpha \\ Q_2 & q \end{bmatrix} A^T = \begin{bmatrix} R \\ 0 \\ 0 \end{bmatrix} \tag{6.2.3}$$

where $[Q_1 : s]$ consists of the first t rows of Q^T, $[Q_2 : s]$ consists of the last $m-t-1$

rows of Q^T. Row $t+1$ can be rotated into rows $t+2,...,m$ to yield

$$\begin{bmatrix} Q_1 & s \\ \tilde{u}^T & \tilde{\alpha} \\ \overline{Q}_2 & 0 \end{bmatrix} A^T = \begin{bmatrix} R \\ 0 \\ 0 \end{bmatrix} \tag{6.2.4}$$

Finally, a sequence of Givens transformations is applied, using row $t+1$ as the pivot row, to zero s:

$$\begin{bmatrix} \overline{Q}_1 & 0 \\ \overline{u}^T & \overline{\alpha} \\ \overline{Q}_2 & 0 \end{bmatrix} \begin{bmatrix} \overline{A}^T \\ v^T \end{bmatrix} = \begin{bmatrix} \overline{R} \\ \overline{w}^T \\ 0 \end{bmatrix} \tag{6.2.5}$$

But since the matrix on the left is orthogonal, it follows that $\overline{\alpha} = 1$ and hence $\overline{u}^T = 0$, $\overline{w} = \overset{+}{-} v$. Consequently,

$$\overline{Q}^T \overline{A}^T = \overline{R} \quad where \quad \overline{Q}^T = \begin{pmatrix} \overline{Q}_1 \\ \overline{Q}_2 \end{pmatrix}$$

and $\overline{Q}$ is orthogonal.

Now let us assume that only Z and R are explicitly available. Partition Z as above:

$$Z^T = \begin{bmatrix} u^T & \alpha \\ Q_2 & q \end{bmatrix}$$

Clearly the first step used above can be applied here to yield

$$\begin{bmatrix} \tilde{u}^T & \tilde{\alpha} \\ \overline{Q}_2 & 0 \end{bmatrix} \begin{bmatrix} \overline{A}^T \\ v^T \end{bmatrix} = \begin{pmatrix} R \\ 0 \end{pmatrix}$$

However, from (6.2.3) we have $R^T s = v$ and so s is computable. A sequence of rotations, G, can now be applied to the matrix $\begin{bmatrix} s & R \\ \tilde{\alpha} & 0 \end{bmatrix}$ with $(\tilde{\alpha},0)$ as the pivot:

$$G \begin{bmatrix} s & R \\ \tilde{\alpha} & 0 \end{bmatrix} = \begin{bmatrix} 0 & \overline{R} \\ 1 & v^T \end{bmatrix}$$

The new orthogonal basis is $\overline{Q}_2$ and the new upper triangular factor is $\overline{R}$. A similar technique can be used for the case when Q is stored in product form.

Updating techniques can also be applied to the elimination method. Indeed we have already considered one updating technique in some detail in Section 3 on linear programming. Another possibility, which avoids sparse matrix updating was suggested by Bisschop and Meeraus[1977,1980].

Suppose that the original basis is $A_0 = L_0 U_0$ and after some time, A_k is defined by replacing columns $e_{i_1}, \ldots, e_{i_k}$ of A_0 with columns $v_1, \ldots, v_k$. The system $A_k x = z$ is equivalent to

$$\begin{bmatrix} A_0 & V_k \\ I_k & 0 \end{bmatrix} \begin{pmatrix} x \\ y \end{pmatrix} = \begin{pmatrix} z \\ 0 \end{pmatrix} \tag{6.2.5}$$

where row j of I_k is e_{i_j} and column j of V_k is v_j. System (6.2.5) can be solved using the factorization

$$\begin{pmatrix} A_0 & V_k \\ I_k & 0 \end{pmatrix} = \begin{pmatrix} A_0 & 0 \\ I_k & C_k \end{pmatrix} \begin{pmatrix} I & A_0^{-1}V_k \\ 0 & I \end{pmatrix}$$

(6.2.6)

where $C_k = -I_k A_0^{-1} V_k$. A factorization of C_k can be updated since modifications involve row and column additions and deletions. (It seems to be necessary to save $A_0^{-1}V_k$) This approach can be continued until it is deemed best to restart.

Matrix Modifications for Range Space Methods

Consider the range space methods discussed in Section 6.1.1. The matrix of concern is M_k:

$$M_k = \begin{pmatrix} B_k & A_k^T \\ A_k & 0 \end{pmatrix}$$

Assume that B is positive definite and so every submatrix B_k is positive definite also.

One possibility is to update the Cholesky factorizations of B_k and C_k. (Is it possible to order the variables and allocate storage for the Cholesky factor of the full sparse matrix B *initially* and then use this fixed space throughout?) The next task is to determine the new Cholesky factor $\bar{L}$, though it is not clear that it is possible to update $\bar{L}$ when a factorization of A is not available (ie. C is obtained via step 2 of the Factor Step.)

Another possibility is to begin with an initial factorization $B_0 = L_0 L_0^T$ and then border the matrix M_0 in a symmetric fashion to reflect subsequent basis changes. This is a symmetric version of the Bisschop-Meeraus[1977,1980] technique described above - we leave the details to the reader.

6.3 Concluding Remarks

We have only touched upon some of the issues involved in large sparse quadratic programming. The next version of this course (1985?) will begin with a thorough discussion of large sparse quadratic programming to be followed with a general nonlinear objective function subject to general nonlinear constraints. A full understanding of the large sparse QP situation is a crucial prequisite to the general problem since a local quadratic model will typically be the mode of attack.

6.4 References

Bisschop, J. and Meeraus, A. [1977]. *Matrix Augmentation and partitioning in the updating of the basis inverse*, Mathematical Programming 13, 241-254.

Bisschop, J. and Meeraus, A. [1980]. *Matrix augmentation and structure preservation in linearly constrained control problems*, Mathematical Programming 18, 7-15.

Fletcher[1981]. *Practical Methods of Optimization: Constrained Optimization*, John Wiley & Sons.

Gill, Golub, Murray, and Saunders [1974]. *Methods for modifying matrix factorizations*, Math. of Computations 28, 505-535.

Gill, Murray, and Wright [1981]. *Practical Optimization*, Academic Press.

Gill, Murray, Wright and Saunders[1983]. *Sparse matrix methods in optimization*, SIAM Journal on Scientific and Statistical Computing, to appear.

George, J.A. and Ng, E. [1983]. *On row and column orderings for sparse least squares problems*, SIAM Journal on Numerical Analysis, 20, 326-344.

Lawson, C. and Hanson, R. [1974]. *Solving Least Squares Problems*, Prentice-Hall.

ol. 117: Fundamentals of Computation Theory. Proceedings, 1981. dited by F. Gécseg. XI, 471 pages. 1981.

ol. 118: Mathematical Foundations of Computer Science 1981. roceedings, 1981. Edited by J. Gruska and M. Chytil. XI, 589 pages. 981.

ol. 119: G. Hirst, Anaphora in Natural Language Understanding: Survey. XIII, 128 pages. 1981.

ol. 120: L. B. Rall, Automatic Differentiation: Techniques and Appli-ations. VIII, 165 pages. 1981.

ol. 121: Z. Zlatev, J. Wasniewski, and K. Schaumburg, Y12M So-tion of Large and Sparse Systems of Linear Algebraic Equations. ∴, 128 pages. 1981.

ol. 122: Algorithms in Modern Mathematics and Computer Science. roceedings, 1979. Edited by A. P. Ershov and D. E. Knuth. XI, 487 ages. 1981.

ol. 123: Trends in Information Processing Systems. Proceedings, 381. Edited by A. J. W. Duijvestijn and P. C. Lockemann. XI, 349 ages. 1981.

ol. 124: W. Polak, Compiler Specification and Verification. XIII, 69 pages. 1981.

ol. 125: Logic of Programs. Proceedings, 1979. Edited by E. ∷ngeler. V, 245 pages. 1981.

ol. 126: Microcomputer System Design. Proceedings, 1981. Edited y M. J. Flynn, N. R. Harris, and D. P. McCarthy. VII, 397 pages. 1982.

oll. 127: Y.Wallach, Alternating Sequential/Parallel Processing. ⟨, 329 pages. 1982.

ol. 128: P. Branquart, G. Louis, P. Wodon, An Analytical Descrip-ion of CHILL, the CCITT High Level Language. VI, 277 pages. 1982.

ol. 129: B. T. Hailpern, Verifying Concurrent Processes Using Temporal Logic. VIII, 208 pages. 1982.

ol. 130: R. Goldblatt, Axiomatising the Logic of Computer Program-ming. XI, 304 pages. 1982.

ol. 131: Logics of Programs. Proceedings, 1981. Edited by D. Kozen. VI, 429 pages. 1982.

ol. 132: Data Base Design Techniques I: Requirements and Logical Structures. Proceedings, 1978. Edited by S.B. Yao, S.B. Navathe, J.L. Weldon, and T.L. Kunii. V, 227 pages. 1982.

Vol. 133: Data Base Design Techniques II: Proceedings, 1979. Edited by S.B. Yao and T.L. Kunii. V, 229–399 pages. 1982.

Vol. 134: Program Specification. Proceedings, 1981. Edited by J. Staunstrup. IV, 426 pages. 1982.

Vol. 135: R.L. Constable, S.D. Johnson, and C.D. Eichenlaub, An Introduction to the PL/CV2 Programming Logic. X, 292 pages. 1982.

Vol. 136: Ch. M. Hoffmann, Group-Theoretic Algorithms and Graph Isomorphism. VIII, 311 pages. 1982.

Vol. 137: International Symposium on Programming. Proceedings, 1982. Edited by M. Dezani-Ciancaglini and M. Montanari. VI, 406 pages. 1982.

Vol. 138: 6th Conference on Automated Deduction. Proceedings, 1982. Edited by D.W. Loveland. VII, 389 pages. 1982.

Vol. 139: J. Uhl, S. Drossopoulou, G. Persch, G. Goos, M. Dausmann, G. Winterstein, W. Kirchgässner, An Attribute Grammar for the Semantic Analysis of Ada. IX, 511 pages. 1982.

Vol. 140: Automata, Languages and programming. Edited by M. Niel-sen and E.M. Schmidt. VII, 614 pages. 1982.

Vol. 141: U. Kastens, B. Hutt, E. Zimmermann, GAG: A Practical Compiler Generator. IV, 156 pages. 1982.

Vol. 142: Problems and Methodologies in Mathematical Software Production. Proceedings, 1980. Edited by P.C. Messina and A. Murli. VII, 271 pages. 1982.

Vol. 143: Operating Systems Engineering. Proceedings, 1980. Edited by M. Maekawa and L.A. Belady. VII, 465 pages. 1982.

Vol. 144: Computer Algebra. Proceedings, 1982. Edited by J. Calmet. XIV, 301 pages. 1982.

Vol. 145: Theoretical Computer Science. Proceedings, 1983. Edited by A.B. Cremers and H.P. Kriegel. X, 367 pages. 1982.

Vol. 146: Research and Development in Information Retrieval. Proceedings, 1982. Edited by G. Salton and H.-J. Schneider. IX, 311 pages. 1983.

Vol. 147: RIMS Symposia on Software Science and Engineering. Proceedings, 1982. Edited by E. Goto, I. Nakata, K. Furukawa, R. Nakajima, and A. Yonezawa. V. 232 pages. 1983.

Vol. 148: Logics of Programs and Their Applications. Proceedings, 1980. Edited by A. Salwicki. VI, 324 pages. 1983.

Vol. 149: Cryptography. Proceedings, 1982. Edited by T. Beth. VIII, 402 pages. 1983.

Vol. 150: Enduser Systems and Their Human Factors. Proceedings, 1983. Edited by A. Blaser and M. Zoeppritz. III, 138 pages. 1983.

Vol. 151: R. Piloty, M. Barbacci, D. Borrione, D. Dietmeyer, F. Hill, and P. Skelly, CONLAN Report. XII, 174 pages. 1983.

Vol. 152: Specification and Design of Software Systems. Proceed-ings, 1982. Edited by E. Knuth and E. J. Neuhold. V, 152 pages. 1983.

Vol. 153: Graph-Grammars and Their Application to Computer Science. Proceedings, 1982. Edited by H. Ehrig, M. Nagl, and G. Rozenberg. VII, 452 pages. 1983.

Vol. 154: Automata, Languages and Programming. Proceedings, 1983. Edited by J. Díaz. VIII, 734 pages. 1983.

Vol. 155: The Programming Language Ada. Reference Manual. Approved 17 February 1983. American National Standards Institute, Inc. ANSI/MIL-STD-1815A-1983. IX, 331 pages. 1983.

Vol. 156: M. H. Overmars, The Design of Dynamic Data Structures. VII, 181 pages. 1983.

Vol. 157: O. Østerby, Z. Zlatev, Direct Methods for Sparse Matrices. VIII, 127 pages. 1983.

Vol. 158: Foundations of Computation Theory. Proceedings, 1983. Edited by M. Karpinski, XI, 517 pages. 1983.

Vol. 159: CAAP'83. Proceedings, 1983. Edited by G. Ausiello and M. Protasi. VI, 416 pages. 1983.

Vol. 160: The IOTA Programming System. Edited by R. Nakajima and T. Yuasa. VII, 217 pages. 1983.

Vol. 161: DIANA, An Intermediate Language for Ada. Edited by G. Goos, W.A. Wulf, A. Evans, Jr. and K.J. Butler. VII, 201 pages. 1983.

Vol. 162: Computer Algebra. Proceedings, 1983. Edited by J.A. van Hulzen. XIII, 305 pages. 1983.

Vol. 163: VLSI Engineering. Proceedings. Edited by T.L. Kunii. VIII, 308 pages. 1984.

Vol. 164: Logics of Programs. Proceedings, 1983. Edited by E. Clarke and D. Kozen. VI, 528 pages. 1984.

Vol. 165: T.F. Coleman, Large Sparse Numerical Optimization. V, 105 pages. 1984.